For DANA!

our actions, our children's crisis

Room for Tomorrow

Bruce Ballister

Bruce Ballister

Room for Tomorrow

Cover Credit: NASA Suomi NPP satellite composite, April 23, 2012

ISBN: 978-1-7332571-4-5 Paperback
       978-1-7332571-3-8 Ebook
LCCN: 2019912498

Other books by Bruce Ballister

Dreamland Diaries
  https://tinyurl.com/h5bjxs
Orion's Light
https://tinyurl.com/yaehvr4q
 Welcome to the Zipper Club, (non-fiction)
https://tinyurl.com/uymp6ga
N. O. K.
  https://tinyurl.com/eyz9dvhf

Reader Reviews:

"Most of us like a spoonful of sugar to make the medicine go down—and Bruce Ballister's Room for Tomorrow provides exactly that, concerning the mess we are making of our planet. A fast-moving plot and edgy characters provide the sugar that makes the grim picture of our future in the year 2214 easier to stomach. The novel also offers hope and a scientifically sound, well-researched prescription for what ails us. His novel joins the growing canon of climate fiction, like Margaret Atwood's *Oryx and Crake* and Barbara Kingsolver's *Flight Risk*.

The sci-fi aspect of the novel is their stumbling onto a portal north of San Francisco that opens into another world. The world of 2214. A world so destroyed by environmental degradation, climate change, shortage of resources, and nuclear war that a few characters from the future time travel to 2020 to pilfer resources and knowledge that has disappeared.

The sci-fi aspect of the novel is their stumbling onto a portal north of San Francisco that opens into another world. The world of 2214. A world so destroyed by environmental degradation, climate change, shortage of resources, and nuclear war that a few characters from the future time travel to 2020 to pilfer resources and knowledge that has disappeared.

The time travelers reluctantly let our clever duo, Parker and Carl, see what the future holds. What they learn shakes the very foundations of their lives. They have trouble accepting that "short term financial greed, religious dogmas, and tribal jealousy" will "bring an end to civilization." As they ponder the horrors facing future generations, they are compelled to risk everything—even their lives—to change the outcome.

Donna Meredith, Associate Editor *of Southern Literary Review* and author of *Wet Work* and *Fraccidental Death*

"While *Room for Tomorrow* is both thriller and Sci-Fi, at its heart, the message it sends – the imperative to take action to save our Earth now – can't be overemphasized."

Kim Ross, Executive Director, ReThink Energy Florida

"Wow, what a ride. It just goes deeper and deeper into the unexpected. I know it has an agenda, but as a fiction piece, I was totally hooked by the characters and their solutions. A great bad guy – good guy cli-fi narrative on when doing the right thing means doing the wrong thing. If I say more, there'll be spoilers; read this book."

Anon, name withheld by request

Ballister's new book takes one of the oldest ideas in science fiction and builds an exciting story about working to solve the problems of today. He doesn't preach. He takes the world as it is and shows that change can happen. At the same time, we are given a love story and adventure that is satisfying and believable."

Mark Odom

Dedication

There is a reference section in the back of this book that provides the reader with further resources on a number of topics that arose in the design of this storyline and the subjects that came up while developing the major focal points.

It is to all of those thousands of researchers, activists, volunteers, and public figures who do get it, who have been in the trenches, that I dedicate this book.

Thank you all,
Bruce Ballister

PART 1

1

La Jolla, California – The Present

At 10:45 on Friday morning, Parker Parrish would never have imagined that by Sunday evening everything about her life would be radically changed. This Friday, she sat unable to get on with the morning's work. The Stanhouser divorce had degenerated into one too many despicable clawing for assets. Another sleepless night had robbed her of needed concentration, or even the desire to concentrate.

A fourth or fifth glance out her window said this particular Friday morning was clear. The usual bowl of smog soup, the morning traffic's stewed and thickened gray-brown fumes, had been blown away by a fresh offshore breeze over the Pasadena Mountains. Her head kept swiveling between the latest of her client's husband's list of demands and the abnormally clear views outside. "Christ, if I could only sleep past 3:30 I might have an ounce of give a damn." She shut the file and stood at the sill, scanning the view—the clay tile roofs of the neighborhood to the west, the dark line of the Pacific beyond—and began to formulate an alternate plan for the rest of the day.

The road beckoned. Her case load certainly did not. Over 500 horses of adrenaline and distraction waited below in the garage.

"Meg?" she called toward her paralegal/receptionist's door, ignoring the hated intercom. A few seconds later, Megumi Sakoda, her good friend, former office mate, and recent employee, peeked in.

"Yes?" The question and raised brows held years of understanding, empathy, and sincere concern for Parker's apparent distracted or diseased state.

"Do we still have any coffee up?" Then, as a second thought, "Did anyone bring in any donuts or muffins?"

"Yes, fresh coffee's ready. Oh, and someone brought in some muffin tops from the Coffee Clutch downstairs. They're in the snack room."

Her seventh-floor, three-room suite shared restrooms and a kitchen with two other professional offices — a high-end real estate broker and a manufacturer's site finder. The arrangement offered the congenial feel of a larger office as well as the necessary intimacy for her small private practice. Her former law partner had recently left for a lobbying job in D.C., leaving her with too many divorces, custodies, and contested wills. The thought of having to find a new partner threatened to bring down what was becoming a hopeful mood.

She'd thought family law would be a refuge from the corporate law firm she'd come to hate. She could not wrap her own decency around defending groundwater polluters against CalEPA enforcement actions. But her new cases were just as ugly, only more personal. "What have I become?"

Startled that she's said it out load, she turned to see if Meg was at the door. Her mouth screwed tight in newfound determination to make a change, even if it was only to her own schedule for one day. She sat back at her jumbled desk and scrolled through her laptop's display: incoming emails, the day's relatively light schedule, an appointment-free weekend, and finally, the next week's calendar entries. A wife beater seeking visitation rights, a high-dollar divorce deposition, and dammit!

What do I really HAVE to do before I can call it a day? She reconsidered the muffin tops, then said yes to the coffee and concluded there was nothing pressing in her inbox.

"Meg," she called back through the seldom-closed doorway, "can you scan the new Stanhouser correspondence for me? Send it to the house email."

She made cursory entries in the calendar and finished off the unusually spare inbox. Meg entered and silently set a mug and small carafe on the side table turned coffee bar. As Meg turned to go, Parker called out, "Thanks!"

Wrapping the cup in both hands, she brought her nose to the rim and slowly inhaled the aroma. *Coffee! Sweet elixir of life! Two amenities I could probably not do without, dark roasted coffee and Megumi Sakoda!* A sip first, (it was still hot), then with acclimated lips a longer pull on the creamed and sweetened lifeblood of the overworked. Nagging indecision still taunted her usual 'get-it-done' work ethic.

~ ~ ~

She sat for a moment, indecision lingering for one last round of "should I or not?" The gray parking lot garage wall gave no answers. She pushed start. Tapping the pedal twice, she brought the feline car to life. The low growl of the Jaguar's tuned exhausts laid a bass line to accompany the thin whine of the car's custom turbocharger. The two disparate but intertwined soundtracks rumbled through the parking garage as Parker maneuvered toward the exit.

She always enjoyed emerging from the dank semi-darkness of her office garage into sunlight. She imagined emerging from the birth canal. Now, as the seconds ticked down on the streetlight, she gave the accelerator one last tap. *Grrruuummm!* Parker watched the crossing lane's light go to amber and flexed her fingers on the Italian leather steering wheel. Green! Pedal! The black beast shot across the four-lane boulevard and up the I-5 ramp into comparatively thin morning traffic. *I'm the boss, and I can leave if I damn well want to!* She accelerated through three gears before the end of the ramp and had to brake to slice between a mini-van and a Corolla to access an inner lane.

A tight grin thinned her lips. Top down, she had threaded her long black hair through the clasp at the back of her Dodgers baseball cap and tamed the ends again with a scrunchie. One of her client's kids had called them "hair clouds" but to her, they would always be scrunchies.

Her sleek, black Euro-muscle car passed delivery vans, cross-town shoppers, cross-country travelers, and the ubiquitous semis as it built momentum. With an eye ahead for anything resembling law enforcement, she pressed on. The speedometer needle passed an even one hundred and climbed onward toward one of four remaining tics on the dial. Tight dimples grew on each cheek, the outward expression of her simple joy. After a brief stretch of unusually empty freeway, she slalomed through light traffic. Her usual shopping exit appeared much sooner than expected, and she blew past it at a little over 120 mph. This was going to be a much better way to spend her Friday. Ahead lay five lanes of pure adrenaline.

Scanning ahead on the freeway, Parker could see two lanes of container carriers converging on a tour bus. That would take out the outer lanes, and the smaller two-axle traffic would try to funnel into the inner and high-occupancy lanes. Her head settled back into the neck pad and headrest as she dropped a gear and sped into a closing opening, the last chance to get past the condensing flow of slower vehicles into free space. Unable to completely shake the office and its end-of-week duties, she started to replay the foreshortened morning and soon found herself whizzing past slow movers at twice their speed. A long, sweeping turn to the northwest and the interstate dipped down to cross the San Elijo Lagoon. She passed a small flotilla of late commuters and shoppers and left behind a metal flake pink '63 Malibu low-rider. The Chevy gave a half-hearted attempt at keeping up but soon dropped back, trailing blue smoke. Too bad about the smoke, she thought; it's such a clear day for a change.

Parker also left behind a thin layer of coastal cloud and emerged into bright sunlight. A speck of brilliant red ahead caught her eye. It too seemed to be slicing and dicing the morning traffic as they both moved north at thirty or more miles an hour faster than the average flow of cars. Two exits ahead, a line of long-distance freight haulers merged from the on ramp. As they struggled to gain momentum on an upgrade, the slow movers filled the other three lanes. She burned a scornful glare at a soccer mom in the HOV lane going, *maybe* the speed limit.

What a waste of space! The red dot ahead was also trapped in the thickening mass of rolling metal. With deft lane changes and determined grit, she slowly made progress toward this new target, a kindred spirit, an imaginary moving goal line. Pass the red sports car. She slowed to almost sixty-five as traffic congealed around her. *Trapped. Damn!* She eyed the clear unobstructed expanses of asphalt ahead in the HOV lane. Too expensive, not again. Those tickets could take all the joy out of the day.

Traffic inevitably slowed. Tapping the steering wheel impatiently, she thought back to the childish euphoria she'd felt when she exited her parking garage to escape Friday's mundane, inane legal manipulations. She was good at it, but hell, the high and mighty of San Diego County would have to wait until next week to sue each other. *How long since I've played hooky?*

The morning had passed with few interruptions and no Hail Mary calls rang in from her widening field of clients. Her strategy to save Herbie Stanhauser from debilitating alimony was solid. *We have the pictures!*

Loosening her grip on the steering wheel, she stretched her neck, rolling it to its limits in all directions, and heard several pops of gristle and sinew, the clicks of bone against what? *Is this me getting older? Did it do that when I was twenty?* She remembered draining that cup of coffee and then Meg. *Crap! Meg!* As a part of a now slow-moving pack, a mere black corpuscle in the asphalt arterial, she texted a short message to Meg: [Sorry for bailing. Feeling a need 4 some RnR. CU Monday AM]. Then as she considered Meg's reaction, she added, [take the afternoon and go play with your kid].

She put the phone back in its Wi-Fi cradle and, looking across several lanes, saw the crimson Sportster half a dozen car lengths ahead. How did that happen? Luck of the lane. It has to go your way some days. It just has to. *Man, I'm tired.* She lifted her sunglasses and examined the reflection in the rear-view mirror. No new wrinkles, but a hint of redness from lack of sleep. She rubbed a knuckle into each eye. *What was that dream all about?* If she'd been moving any slower, she'd be tempted to pull over, shut her eyes, and try to reconstruct the dream. That damned dream and its night terrors had lost her so much sleep over the last four, maybe five days!

She hadn't seen a farmhouse like that since her country girl days back home in Florida. A light year away it seemed. Something about a dark hallway and a door? A honk from behind found her, lids shut, embarrassed. An empty lane lay ahead with cars form adjoining lanes trying to fill the gap from each side.

Parker took a deep breath and settled her mind to blow out the weeds, get out of the snarl and get on up the coast to a favorite cliff-top overlook. The smile she'd had back in La Jolla had morphed into a grim pucker, and she was determined to erase it one way or another. A slow-to-react semi to her right gave her the first opening, and as the pack began to approach posted speeds, she took other openings and made a few others that bordered on rude or dangerous. Her distance traveled began to put space between the troubles of the previous night and the present day. She worked back toward the faster-moving inside lanes, and when a Jeep full of jocks passed by in the HOV lane, she took her chance. She slid left behind the Jeep and accelerated. As the bulk of the traffic clot passed by to her right, she noted the rooftop of the red sportster, slowing behind a tanker. It maneuvered into a barely existing gap and began to pull ahead, only to get blocked again. The Jeep didn't seem to be inspired to make time, so she found gaps in the main flow that offered more advantage.

Spaces opened in the pack, and she resumed her favorite high-speed slalom through fairly open traffic. The crimson sportster often disappeared from view because of its low stature and her own weaving, but car length by car length, she drew closer. As her Jag slid between a housewife in a CRV and a Mack pulling a fuel tanker, the red pulled beside her from the other side of the truck. He flashed a brilliant grin and a wave as he passed, a shock of sun-bleached blonde hair blowing wildly. She'd gotten boxed in by the CRV as he made several car length's progress. Traffic in the adjacent lane slowed. She caught red in her peripheral vision as she downshifted and powered through a slot left, the far inside lane. A brief interlude in the HOV lane again, and she cleared the platoon of slow movers.

Parker's day was improving. The grin retaking her face had grit and steel. In too few seconds, her Jag caught up to the next clump of slow movers. A quarter mile back, a speck of red emerged from the pack shimmering in heat waves coming off the freeway. It accelerated toward her at, she estimated, well over a hundred. In a few seconds, the whine of her Jag's turbocharged breathers locked tempo with his as the crimson Porsche slowed into the lane beside her. Again, the flash of teeth, and now a nod. She followed his glance forward and saw a gap just before the Porsche leapt into it. Again, black trailed red.

They made ten miles further in this manner, swapping off the lead, blocking for each other, developing a rhythm, a trust bond. Another wall of vehicles three or four deep slowed their dance to a pedestrian seventy-five. A subtle puff of black from the chrome tower exhausts of a bob-tailed Fruehauf clued her to a shift in the wall of blocking vehicles, and she was ready for the opening. She took the lead again, flashing a brake light to force a commuter to brake. The Porsche followed her across three lanes in a stair-step advance toward the outside where forward progress slowed again.

The red Porsche passed on the far-right shoulder and, braking back into the line, opened a hole for her to get in. At another blockage, they took additional liberties with the break-down lane to pass a quarter mile-long, five-lane, slow-moving parking lot. Together, the two dodged crazily through late morning traffic. Twenty miles disappeared in less than fifteen minutes. With an abrupt signal and a hand wave, she took a ramp and pulled off to a local street that led under the freeway, through a posh residential area and down to an ocean overlook. The Porsche followed, pulling in beside her. She shut down her car, looked over at the Porsche. Its driver's door cracked but did not fully open.

~ ~ ~

Their transient bond, fired by adrenaline and gasoline, required a meet-up. Who was this other spirit? She put aside her usually constrained manner and determined to meet him. The two drivers' doors both opened as if on cue. A smile and that wild thatch of wind-blown golden hair! A younger man, well, not too much younger, swiveled out of the Porsche's driver's seat and stood, returning her smug grin. *How did a guy that tall sandwich into a cockpit that small?*

"Hello, I'm Parker." She extended her hand.

The younger man, matching her appraising gaze, took her hand. "Hi, Mrs. Parker. I'm Carl, er, Carlos, Carlos Reyes. Everyone calls me Carl."

She swallowed an abashed grin, straightened her mouth, and corrected him. "I'm Parker Parrish, and it's not a Missus and certainly not Miss. Just Parker, Parker Parrish." *God, I hope he's not still in college.* She nodded to his car. "Nice restore job on your car. Is that a 930?"

"Yeah, it's an '85, slant nose." She noted the loving look on his face as he turned momentarily to match her line of sight.

7

He turned back to her, questions in his eyes. "How'd you know it's a restore?"

"Just a guess. By now most of those have been bashed in one corner or another, and most have rebuilt or modified engines. Besides, that looks like new leather." She scanned the rolled red and white upholstery and admired the carved burl dashboard. When she looked up, he had to quickly adjust his eyes from her legs.

Parker was a little amused at his unabashed, quick survey of her personal details. She took stock of her presentation: business suit, not too short, but enough leg to judge her athleticism. Her hair had lost the lower tie and now blossomed from the back of her baseball cap. She hoped her usually clear blue eyes didn't betray lack of sleep as she maintained his stare. Others before Carl had commented on her narrow, straight nose, small mouth, and defined cheekbones. She had no lipstick this morning on lips that curled now into a sideways smile. She had fair skin that needed a little more tan to blend with the Southern California natives.

He seemed to like what he saw, and after only the briefest pause for the assessment, said, "Nice to know you. Nice to know you like good cars." Then, examining her car, "Nice Jag. That an XKR? 2015?"

She shrugged. "Yes, I'm afraid it's one of the last or its breed. XKR-S, supercharged, 550 horses. Top speed is unattainable on these freeways. Nice perk for a youth misspent doing my best to be the best." She wondered that he knew so much about the line-up of Jaguars.

"Yeah? What are you best at?"

"I didn't say I *was* the best, just that I tried very hard at it."

"Well," he waved a hand at the Jaguar, "you seemed to have gotten very good at something."

"Let's just say that if you ever got in trouble with that big-boy toy, you might need to give me a call." She leaned in to get her purse and fished out the gold case that kept her special cards, the ones that included her mobile number. "Call me if they say you only get one phone call."

He took the formal, white, heavy card with black serif lettering, reading: "Parker Parrish, Attorney at Law. Parrish and Cole, Family Law."

"At your service." She extended her hand again, this time a little more formally. "Actually, I'm on my own now, but that number is still good. And I still have a few corporate clients. Much more money!"

Carl took the offered hand and she felt firmness in the grip. So many guys gave women a wimpy handshake. He asked, "So where were you going in such a hurry?"

"Absolutely nowhere in particular. I was a little tired and decided to play hooky. The fresh air perked me up, and the pedal took over." She thought back to one of their riskier maneuvers. "I saw you trying to get around that tanker truck. I was almost afraid you'd try to go under it to get past the mini-van."

"Yeah, well, thank you for pointing out that a paved shoulder is just as good as a fifth lane." He made a show of bowing in tribute. "So, like, where are you going now?"

"So, like," delivered in mockery of the overused interjection, "I don't have any idea. I think before we had our little adrenaline-fired interlude, I was heading here just to gaze out to sea." She nodded toward a path that led away from the parking area and to a slightly higher overlook. "I was actually looking forward to doing that alone. Hope you don't mind."

Parker saw physical deflation on the younger man's face.

He said. "Sure. See ya." The smile in his eyes was genuine, but his mouth showed disappointment. As he turned toward the Porsche, she was suddenly aware of the near quiet of the place. The overlook was shielded from the freeway noise above, leaving the surf sound from the bottom of the cliff to susurrate in the void as he walked away.

When Carl reached his car, she added, "Drive safely, OK? Seriously, I'd hate to read about your tragic death."

Hand on the door handle, he recovered quickly. "No problem. I was just in the mood when you went by. Glad I was able to keep up." He nodded appreciatively, a small grin spreading. "Good run. If you ever want to go the other way, take them out in the desert and air them out, blow the dust off the carbs, let me know."

"How would I do that? Card?" She wondered if she'd misspoken. Did he have a card? Was Carl still in college?

Returning, he pulled out his wallet and fished for a bright blue, glossy plasticized card. She read "Carlos Reyes, SVP, Tech

Services, So-Called-Wizards, Inc." She considered a mental list of acronyms. "Senior Vice President?"

"Yes, actually though I'm really a super geek. I guess I'm pretty good at what I do too, and," he paused, "the number's still good, but I'm on my own now."

Parker waited, drawing him out with a slow raised-eyebrow nod. "You guess?" She usually wasn't challenged by mid-level talent. "It says you're only a So-Called wizard."

He tilted his head, grinning. "That's for Southern California Wizards. Let's just say that if you get hacked, or even think you might be hacked, I'm the one you should call." He smiled, waved a hand in the air. "I'm a pro at corporate security."

"I hope it doesn't come to that. I mean, not that I wouldn't mind taking the cars out for a spin-up. I hope I don't need your tech services." She looked meaningfully at the trail into the needlegrass.

"Gotcha." He was making an honorable exit. "Hope to see your Jag in traffic again sometime."

"You, too. See ya." She watched as he squirmed his light muscular frame into the little Porsche. *Well, he is slightly older than I first thought. And cute, and...*She let the thought drop and turned to the trail. She turned abruptly, back toward the younger man in the brilliant carmine Porsche. Something had clicked between them. Maybe, maybe. "Carl?"

The door opened and after only a slight hesitation, Carl got out again. Smiling. She wondered, *If I can tell this story, not too incoherently, maybe I can put a stop to the damn dreams.* He was here now, the sea breeze tugging gently on his unruly hair. "Carl, I'd like to run something by you, see if I sound crazy. Rather, see if this scenario sounds crazy."

"Sure, I'm already about thirty miles north of where I was going anyway." He followed her to a cliff-edge viewing overlook and stood beside her. Shading his eyes against the glare, he peered out across the dried grasses to the white crested breakers crashing below and waited.

Where to start? "Carl, do you ever have vivid dreams? Dreams you remember?

He paused, brows knitting. "Sometimes," he said, watching the train of approaching swells. She noticed his eye catch the path of a diving gull. Facing her again, he added, "But mostly they're gone by the time I'm pouring coffee. Even the best ones seem to fade if I don't wonder about them, try to seek deeper meaning." He looked changed; the intimate question seemed to add a few years to her estimate of his age. He'd looked back out across a low stone wall at the sun-glinted waves.

She waited for him to look back to her and continued. "The thing is, there's this dream I've been having. It usually hits about three a.m. I wake in a cold sweat. I've either tossed all my sheets off or I'm tangled up in them."

"Jeez, and they're repeating?" His face had transformed into the picture of empathy. "For how long?" He leaned in, clearly trying to listen, to show that he was listening.

"'Bout a week, I think. And yes, that's half the reason I'm out of the office today, ignoring my to-do list." She absently picked at a rumpled candy wrapper and folded it gooey side in. She looked back up at Carl. Pointing at an overlook bench, she said, "Let's have a seat. It might take a little while."

~ ~ ~

At 3:27 Saturday morning, Parker flashed awake again. Her breathing was shallow and rapid. The heartbeat in her ears pounded out a near-panic tempo. Once again, the dream had taken her to the brink — to the end of the hall — and narrowed her scope of vision to the faux crystal doorknob, even lowered it to the keyhole below, and then the dream pushed her out, refusing any further progress.

2

She took a deep breath to slow her panting and collect her thoughts. She needed to rearrange her reality almost every time she came out of it. How many times now? Six? Seven? Getting back to her bedroom and familiar surroundings always felt like coming back from some other physical place and from some other point in time. She'd had overly real and disorienting dreams before, but they had not given her the dynamic tension of both dread and attraction.

Jazzman's squirming body, desperate to avoid being crushed, halted her attempt to roll out of her unusually uncomfortable sleeping position. "Sorry, Jazz." Parker waited for her cat to get free and stretched out flat on her back. *DAMN! I wish this dream would either let me in the door or stop bothering me. Or just stop.* The clock clicked, and 3:28 cast its lime green LED glow into the room. She turned her head to the clock and acknowledged the continued pattern. Just after three again. *This is no way to catch up on last night's lost sleep.* Her cat, its sleep momentarily interrupted, circled twice and curled into the recess between her arm and body.

Damn dream. If I go back in on purpose, maybe I can resolve this thing and get some rest. Out of desperation to make sense of the final scene, she put herself again at the front of the house and tried to retrace how the dream always brought her in. Fully awake, she centered her mind's eye in the overgrown red brick path to visualize the house. The clapboard farmhouse sags on moss-covered piers. It's partially shielded by overgrown shrubs at the foundation and high, wild grasses and weeds in the wreck of a yard. An overgrown brick path leads to the porch steps. A barely discernible trail, a mere thinning of the overgrowth, yet too well traveled to be a rabbit trail, leads around the right side of the house.

Two double-sash windows flank each side of the central front door made of a dark wood. Six glass panes in the door expose faded and tattered sheers. Blistered and peeling white paint hangs limply from the frame. A layer of dust and grime covers any horizontal edges on the lacquer-glazed door. Dust from decades of neglect is glued to the remains of a yellowing paint job. The porch wood, old tongue-and-groove pine, is solid enough, just filthy with windblown dirt and curled leaves. Empty pints of liquor, crushed beer cans, and snack wrappers litter the porch and yard. Above the porch roof, a bank of three windows looks out over the front yard. All in all, she remembered the outside of the house looking as if no one had given it any loving attention for far too long. When the roof failed, the house would go. Would anyone besides lost hikers miss it? She took up her visual tour again, from the inside.

The brass doorknob is dull; no hands had polished it in a long time. But wait, the dream never took her into the house through the front door. She had always found her way in through a side entrance to the kitchen. A kitchen that had never been remodeled still had a chipped porcelain worktable and open shelving above a faded Formica counter. She focused on the thin track through the side yard to the back.

Somewhere in the pre-dawn dark, brakes squealed, followed by a low diesel growl. Consciousness intruded on the rustic farmhouse scene again. Her attempt to recreate the dream was shattered by the juxtaposition of its isolated country farmhouse setting and the urban interruption of the recycling trucks' early-morning rounds. The far-off duet of grumble and screech announced the waste management truck's progression through the neighborhood.

She was in her own reality again. Jazzman's purring thrummed through her own ribs. A glance said it was 3:34. She gently nudged the cat away, barely conscious of its leaving. She took a deep breath and, with closed lids, went back to rebuild the setting.

Why were the trees always barren? Was it winter? Were they dead? Parker backed her mind's eye away from the house and tried to see more of what was on the periphery. More dead or wintering trees. The setting was on a street. No, a road. Certainly not in a suburban neighborhood. She could not turn around to see what lay around behind her or what the view was in either direction along the road. The house seemed to want to be in the center of her view. It asked her in, drew her in, called to every inch of her being, and like a frustrated lover, refused to let her take the last step, refused to let her into the beguiling room at the end of the upstairs hallway.

Her mind moved along the overgrown, bricked path again. Turning right, she followed a rabbit trail through the high grasses of the front and side yards. *Odd, the grass makes no sound as I move through it. Am I dreaming, or is this me forcing it?*

There are the few steps leading to the kitchen door and the peeling gray paint that remains in a few places on the poured concrete steps. Her feet rise carefully onto the first two steps before her viewpoint moves to the four inches of slightly opened doorway. Beyond, the oak worktable shows stains, cuts, and a few burn marks from decades of use as the room's primary work surface. Her hand pulls open the door.

No squeaks come from the ancient hinges. In the movies there would be a squeak. Something moves furtively in a corner and disappears before she can locate the noise. She notices minute tracks in the dust-covered floor. *Why wouldn't there be a house mouse?*

Her eyes popped open. Too analytical.

She tried again to re-enter the dream. Front door, again at the side door. Is there a back door? Nothing would come of it. A half hour later, nearing exhaustion, her body finally took over and let her find sleep.

Saturday dawned slowly. She found herself again tangled, foot in sheet, groggy and exhausted. *What the hell? I should just get up. How many times should I put off...*her cat leapt from its nocturnal perch on the shelves beside the bed onto her pillow. Parker jerked away, startled. "Good morning to you too, Jazzman." Her cat, black except for a thin white diamond blaze on his chest, nuzzled his head into her neck and thrummed a hello.

She relaxed again, closed her eyelids, and focused on the path around to the kitchen door. She was almost in, back in dream-space when Jazzman's subtle pressures on the mattress moved closer. His contented purring, agreeably familiar, yet distractingly close, ended all chances of getting back into the house, or even back to sleep. She reached into the dark and pulled him close.

"Bastard!" She rubbed his neck roughly, but he pressed harder into her fingers. "Hedonist!" In exhausted desperation, she closed her eyes.

Parker woke sometime later thanks to her early morning efforts to drop her almost total blackout shades. She headed down the hall toward the stairs. No physical blocks this time. No reflexive inhibitions as she passed the doors to her reading room or the guest bedroom. Only the scampering black form of Jazzman running ahead to get to the cat food bin.

His usual routine was to beat her to the bottom of the stairs, dash to the kitchen counter, and leap onto the top of the plastic bin that stored a weekly supply of his dry food. In this position, his repetitive meowing was anything but subtle. He usually got his reward as her first function in the kitchen, right after she hit the button on the coffeemaker.

This Saturday morning, Parker stopped. Something in her memory flashed. A jarring picture had popped unbidden to the fore and stopped her in her tracks. Complaining or not, Jazzman would have to wait. She sat on the turn at the bottom of the stairs, three steps up from the first floor, and put her head in her hands, forcing blackness on her eyelids.

There's something about the house. This time, it's a wider scene. The house is to her right and she's approaching slowly. A wavering motion jars the scene. She's walking; that's the unusual motion. And the sky. In this scene, brilliant daylight in the God-given clearness of a smog-free day lights the scene. The trees are still barren, blackened by brushfire. The grasses nearest the house have somehow been spared from being scorched, but the hillside that wraps around the house is a burn-blackened scar. She turns to see her Jeep parked a few hundred feet up the road and downhill, a sunlit valley, checker-boarded by farms, spreads out—yellow, tan, and green beyond.

The valley images vanished, drawing her back to the staircase. But the image was sharp, more memory than imagination, lacking the surreal lighting and indistinct tunnel vision of a dream state. *Could this be memory or…or madness?* She shivered as something touched her arm, her shoulder, her face. *Oh God!* Cries for help invaded her space. Then Jazzman licked her face. Damned cat!

~ ~ ~

Showered and breakfasted, Parker decided to leave, to go for that drive in the desert hinted at by … who was it, Carlos? Carl? Rummaging in her bag, she found his card. Texting, she gave him directions and a time. Then, looking at her phone's map app, she provided a GPS coordinate pair to a valley overlook and an approximate time to meet on the Ortega Highway northeast of San Juan Capistrano. The drive would allow her to clear her head, force her to pay attention. Anything less than absolute concentration at her planned speeds and she would kill herself or an innocent oncoming driver.

A little over an hour later, Parker pulled off the Ortega Highway onto an overlook. She walked off tightened muscles and leaned back against the parking area's protective stone masonry wall. A cloud deck shrouding the coastal metroplex was visible in the valley below as a downy blanket of white, a purifying blanket of mist that obscured all the major and minor tales of love and loss and deceit in the greater Los Angeles area. With all the cases she handled — divorces, property line disputes, dishonest contractors, severed lives, blatantly unfair corporate scams and swindles — Saturdays were her escape. The Jaguar was her vehicle. The dream, that damned dream, was a distraction, signifying what? Nothing, she hoped.

She had given the dream some thought driving up the 74. At first, the curves had been gentle, fields on one side, hillocks on the other. A small turn to the right and the required level of attention ratcheted up. No more time for reflection. The drive was a challenge at the speed limit. At ninety, it was exhilarating. White knuckle. Breathtaking, gut-grabbing excitement on four wheels. Speed kills, they say, and so did driving too fast. She'd never taken amp, amphies, or blue-rock, fet, whizz, or any of the pet names some of her least-desirable clients called amphetamine, but she enjoyed viscerally pushing the limits of her car's abilities. She'd left the dream behind, below the cloud deck, out of sight and, now that she could see a red dot moving rather too fast up the canyon, out of mind.

Parker slid back into her Jag and, backing, waited to see the Porsche rounding a bend just down the road. There was nothing moving between them. She pulled out, leaving a small blue cloud of singed rubber at the edge of the pavement. The two sped eastwards, the morning sun occasionally flaring the view ahead in painfully intense whites. A slight change of direction and they sped into blue shadows of nearly vertical rock walls. The gentle climb topped out at only 2,700 feet. She pulled off into the next overlook, fishtailing slightly, throwing gravel over the edge.

As she pressed the button to lower the convertible top into the boot, Carl's red P930 pulled in beside her, rattling gravel and blowing a cloud of white dust up between them. "Hey!" she shouted a little too loudly, briefly regretting opening the car up to the settling dust. She noted his chagrin and, wagging her head, smiled her acceptance. Calling out over the noise of the two powerful engines, she asked. "Wanna come with or follow tandem?" He was already out, his engine off. In answer, he hopped in beside her and clicked his key at the Porsche, satisfied by its "eep-eep."

The conversation was a little tenuous at first. How did an older 930 get automatic door locks? Expensive aftermarket anti-theft system. Who did the work? Do they work on Jags? Mundane and unnecessary preliminaries to the things that really matter when two singles are assessing their chances of becoming half of a pair. The trip down the eastern slopes toward Lake Elsinore was steeper and the drive through the desert communities dull, but they had time to share the mini-histories that would normally be shouted over high-top tables for two in one of the hundreds of nightclubs in Orange or Los Angeles Counties. They drove eastward, across desiccated retirement and farming communities at various pegs on the economic scale, past cactus gardens and green-sprayed rock for yards. This was not her kind of California. She needed humidity; she'd been born, raised, and educated in North Florida. Though her cliff-top house on the La Jolla coastline was nothing like her beloved North Florida's pine-covered flatlands, at least it had moisture going for it.

By noon they were racing away from Palm Springs on Interstate 10. Only 2,300 miles to the Atlantic Ocean. In her car, without the restrictions of speed limits, she could be back in her hometown in the Florida panhandle in a little over twenty-four hours, including gas stops and pee breaks. There and back to work by Monday evening. THAT would be a bonding experience! After a brief gas-food-biological break in Indio, she was able to open it up. I-10 was straight enough and nearly flat. The only danger was in passing a random car that might change lanes unexpectedly. The citizens, as she and Carl had begun to call them, were going seventy to eighty. The Jaguar sped past them at over 120.

Her car literally floated over the road. Her worries floated too — too far behind in the back eddies to catch up. She liked Carl. He was complex for thirty-three, which was older than she had originally taken him for. She had thought maybe twenty-fourish, if that, when she first saw him at the Pacific overlook. He was well preserved for an over-caffeinated tech wizard who, she learned, had grown his own company and sold it for millions. He had hired on as a consultant with an hourly charge rate double or triple her own, and by force of talent, made senior vice president. So-Called Wizards: So-Called for Southern California and Wizard was tech-speak for what he did. The rank and file did standard upgrades, server maintenance and the routine drudge work of the tech world. He and his former partner had more extensive and sometimes not quite legal work in IT security. He worked when the job was a challenge because he loved his work. He kept bad people from doing worse things to secure data. Not everyone they worked for was pristine, but their clients wanted the security his form of magic could afford them. She just worked because she was good at it and it afforded her a house with an ocean view where ocean views added a million dollars to the base price of very nice homes.

They stopped at a forgettable cinder block abomination of a motel that featured rooms by the hour, half-day, even all night. They showered together and consummated their first date on sheets they hoped were clean, nervously laughing about that and the tacky cactus desert art. After a fried chicken dinner ("do you want fries or mashed, ma'am?"), they were back on the road, chasing the setting sun. They bumped phones at the overlook and promised to get in touch. She wondered if they would. What would a slightly over-forty lawyer lady and an almost decade-younger tech wiz have in common besides their shared love of velocity? She had not spoken further of her dreams. But the dream; they had talked about pasts and futures, but not dreams. Not *the* dream.

And she wasn't at all disappointed with some of her early relationship digging. While lying exhausted, legs and limbs still entangled, she had plied him with her basic "new guy" questionnaire.

"Have you ever been married?"

"Once. So over."

"Have you ever had any other long-term meaningful relationship?"

After a pause, "Yes, one."

"Have you ever cheated on either relationship?"

"No, with my thoughts maybe, but not in any overt action."

He'd passed. The relationship had come to ground. She'd thought back on the few relationships she'd had that grew to something serious. The sex had to be good, grounding, establishing a floor on which everything else could be built. She and Carl had, in only a few hours, established that trust bond, the grounding she would soon find she needed.

3

3:16 Sunday morning. Parker woke with a start, her breath held in anticipation of some unknowable danger. She had been at the doorway to "the room," unable to reach forward to grasp the knob. Urging her hand forward could not make it so. In the silent darkness of her room, she could hear Jazzman's comforting snore. She rolled out of bed to make an obligatory rest stop and decided a change of pace would be good for her head. Still tired from the prior day's desert trip and the quick tryst in the no-tell motel, she went to her reading room and looked for anything she could find on dreams. Nothing relevant presented itself in her library.

Pulling out a seldom used iPad, she scrolled through several hits on "recurring dreams," most of which seemed to be fronts for some savant's book or a miracle medicine. Frustrated, she leaned back into the rolls of heavy chenille upholstery. Her wingback chair was inherited from her grandmother's house in Florida, and it had always seemed like the chair provided needed hugs when nothing else or no one was available. Jazzman followed her into the chair and curled into a black ball with ears. She settled into the chair's tucked and rolled overstuffed pleats and slept hard until dawn.

~ ~ ~

Fortified with breakfast and caffeine, she decided on the Jeep, brushed two weeks' accumulated leaf litter from the aging Cherokee's AC vents, and headed north. She had intended to simply drive up the Pacific Coast Highway, get through LA before it had a chance to wake up, and keep going. Fog-bound cliffs changed her plan and she took the 405 through LA County, and two hours later found herself pulling off at the Spanish Point lookout over Pyramid Lake. She walked uncertainly toward the lake, the natural point of interest at the overlook, and smiled as a flock of water birds floated in, soaring on unmoving wings till the last; to a noisy quacking splashdown. Scraggly pines, burnt and blackened, stood in ranks of ragged cuneiform along a ridge across the lake. The blackened hillside testified to last summer's fires. Her side of the lake had escaped the blazes and was redolent with pine and juniper. With scant traffic on the freeway, she found a certain closed-eyelid peace with a nose full of pine. Breathing the scents in deeply, she noted that these western pines smelled altogether different from the familiar slash pine plantations back in Florida.

Opening her eyes, she scanned the full horizon and noted a roadway — no, a dirt track — with multiple switchbacks leading uphill. Jeep trail! The two-rut track was on the other side of the interstate. A lone semi drifted downgrade, softly exhaling a throaty growl as the driver engine braked in advance of a steeper grade. She followed its progress as it passed over what initially appeared to be a short bridge but which turned out to be a park traffic access tunnel under the four-lane. A quick scan of the roadside park's kiosk verified that she could get to that trail from here.

Planless still but determined to find some adventure to quench her preoccupation with the damned farmhouse, Parker took the tunnel and found a gated entrance to the dirt road she'd seen from the overlook. She sighed, disappointed, looking beyond the gate at the inviting dirt drive, and then refocused on the gate. The lock holding the two ends of the simple chain was in place but not locked. No tracks were visible in the sandy portions of the road, despite the long absence of rain. This was a singularly private road.

With casual but certain motions, she moved to the gate, loosed the chain, and let the corrugated metal frame swing open. She was through and gone, with the feckless lock and chain in place behind her, in less than two minutes. She noted a slight smoky tinge to the air; the morning damp enhanced the smell of charred brush. Above the gate and around two turns, the two-rut trail improved to a graded road and greener ground cover. She took curves at that edge of controlled instability where any more speed would send her into a wall or chasm. Four-wheeler paradise. A grader had been in this section since the last rain a month ago, leaving the surface free of washboarding but surfaced by a fine gravel/sand mix. It fueled her adrenaline need just fine. At the top of the grade there was a T-intersection, and flipping a mental coin, she took the left branch.

Going down slope was a little hairier, and she let her engine do most of the braking when needed. It would not do to lock up the wheels. Halfway down, she spotted the agricultural flatlands south of Bakersfield and knew with some certainty where she was. A sense, not quite of déjà vu, but of actually having been there before settled in. She passed into another area that had burned within the past few years. Blackened scrub, leafless trees, and cinders covered most slopes. In a few areas, specks of green poked between rock and crevice.

A turn to the right near the bottom brought her to aging pavement at another tee. Another mental coin flip took her to the right. She had barely accelerated when an overwhelming sense of dread threw a pall on the morning. She rolled to a stop on the shoulder and felt her pulse racing. What? She scanned the valley below; it was familiar, but then, it didn't look much different than it might from the freeway overlooking Bakersfield. What was here? For that matter, why was she here? Unplanned turns left or right; random choices, she'd thought. She shut her eyes, curled her hand in her lap, and ran through a few sequences of yoga breathing. Hatha had not done much for her body tone, but it had taught her how to calm her beating heart just before delivering a knockout argument to a district attorney.

When she opened them, it was noticeably darker outside. The cloud deck on the Pacific side was piling over the ridgeline, sucking the joy from the morning sky. Looking up, she was amazed at how fast the weather front moved in and, tucked into the contours of the mountain as she was, how close it seemed. The first raindrop splatted right in front of her face, startling her. The second hit her elbow, which she pulled in to roll up the window. As she sat and watched over the next two minutes, a light shower fell, dampening the road ahead as well as her formerly high spirits.

She took one last breath and turned the ignition. She felt the familiar roar of the Jeep's fan as it revved. Why couldn't American engineers figure out how to make an efficient fan? Germans, French, Swedes, and hell, even the Italians could make a quiet engine fan. Jeep? Not so much. But then, neither the Germans nor the Swedes could make a Jeep. As she sat, wipers off, droplets filled the windshield and began to run in tiny rivulets. The narrow park road ahead was dirt and getting muddier by the minute. Parker slowly pulled back into the narrow lane and sped up to the park's speed limit of twenty-five. The Jeep rounded one more turn into a canyon and the chill returned. She was reaching for the car's heater controls when she slammed the brakes and slid to a stop.

Son of a bitch! Ahead, tucked into a curve, was the house! Her pulse had spiked again. Her breathing, stopped. The house was set into a hollow with burnt mountain slopes on three sides. She released the brake and let the car roll slowly toward the front yard of THE house.

~ ~ ~

There was no question. For a moment, she just stared and let the detail of the real thing paint over the dream's reconstruction. The odd smell of the dream was not dissimilar to the damp carbon smell of fresh rain on the scorched underbrush. These trees would be barren no matter what season. They would never again bear leaves. Firemen must have laid down a tremendous amount of water to keep the house from being taken. Why?

With the shower tapering to a light drizzle, it was easy to see the house was an unoccupied wreck. A little garage/storage building to the rear had been singed but was almost where the dream had placed it. Odd that a double row of tire tracks laid down in broken grass led to its door. The front yard was a tall stand of needlegrass, overgrown and bending down now, heavy with rain. She cracked the Jeep's door, hesitated, and got out, glad for having put on serious cross-trainers instead of some less worthy outdoor shoes when she dressed for the drive. Never knew when she might want to sprint. Red brick pavers still showed a thin path, worn by whom? Small animals? Homeless? Hikers? Druggies? Hell, she could be walking toward a booby-trapped coke kitchen. She could imagine the headline: "Noted Orange County Attorney Killed in Crack House Explosion." Parker felt foolish, apprehensive, and a little frightened.

THIS is not a dream. I can feel mist on my cheeks. Dreams don't do that, do they? She pinched the web between thumb and forefinger. It hurt. There were no sounds other than the light patter of raindrops on the porch's tin roof and the background ssshhh of light rain on a light covering of emerging grass. No movement betrayed any possible occupant sheltering beyond the darkened windows. She approached and remembered to breathe.

The second front step creaked as it always had in every replay. She knew without trying that the front door was locked. She tried anyway. The worn leading edge of the tongue and groove floorboards had all the chips in all the right places. She did not remember that the left side of the porch had one caned chair lying on its side with the caning broken out of the seat. Maybe someone had moved it there since I was here last. When was I here last? Surely, I must have been here. But when? She stepped down off the porch and followed the thin animal trail around to the right. She looked out across the front yard, noting that her Jeep was really just fifty fast steps away.

The familiar peeling, gray-painted concrete steps led to the kitchen door. Two sets of footprints, dappled by the light rain, led toward a low stone wall at the back of the scrubby yard. She examined the door; rusted hinges spoke of a long-gone screen door. The half-glass kitchen door was closed. It had always been ajar in her dream-state tours. Parker grasped the glass doorknob and felt the catch release. She pushed and let the door swing in. No eerie scary-house screech; it was ironically smooth, as if it had been recently oiled. A time-worn oak worktable dominated the small room, with barely enough passing room on either side of the rough and stained wood planks. The table's surface was scarred from ages of food prep. A chipped porcelain sink showed its black iron in too many places around the edges. Open shelving, not cabinets, framed the window that faced the small backyard. Scanning its limits, she noted again the rock wall. She thought it probably had grown every time another stone rolled down the slope from the hillside above.

Parker tried the cold-water handle. After a hiss and spit, and a water hammer knock somewhere under the floor, clear water spilled out, wetting the dust in the sink bottom. She knew she was delaying what she had to do. She turned and faced the dogtrot hallway that ran through the middle of the house to the front door. A small sitting room lay to the right, with a tiny bathroom tucked under stairs, and a living room to the left as dusty and barren as the day the last occupant moved out. She thought this derelict was too far from the traveled highways in the valley to attract many homeless and wondered whether it was in park land or was abandoned private property. The place was relatively clean for an abandoned property. No aging fast food bags or MRE boxes from a casual overnighter or camper looking for relief from weather. No longnecks or pint bottles left behind by the alcoholic homeless. Her forensically trained eye especially noticed the floor. The floor was not a dusty mess. She left no footprints.

The newel post at the base of the stairs was solid after what must have been generations of use, its worn round knob at the top polished by millions of passing hands. She grabbed it and pulled herself up to the first step. She paused, not knowing why, to breathe, to reconsider. She turned back toward the front door, thought of unlocking it to prepare for a fast escape. A glance through tattered gauze window sheers found the reassuring mass of her dark teal Jeep waiting just beyond the overgrown yard. She stepped back, released the slide lock on the sheet metal brass door lock, and pulled the door slightly open. Biting her lower lip, she pulled the door wide open, letting it come to rest against the spring stop in the base board. It would be an easy pass-through now if she had to come down these stairs in a hurry.

Thin rivulets of rain dribbled off the tin porch roof onto the next to last step. She made a note to leap all the way to the path if it came to that. No use slipping on that step and letting, what or who, catch her. She looked up at the car. It would be easy to go now, to slide into its comfortable safety, drive on down to the valley floor, and take the freeway back home to a glass of wine, a book and Jazzman's noisy company. Maybe a quiet Sunday at home with the cat would be the sensible thing. But she knew she could not do the sensible thing. She'd be drawn back here, when, tomorrow? Next week? She'd be back to face…something. Or nothing. If she didn't try now, the dream would be back again at zero-dark-thirty.

Resolved, she turned and took the first three steps up to the corner landing. She laughed a quiet, personal laugh. *Same as Mom's house, three steps to a landing, and stairs up to another landing two steps shy of the top.* These turned left, Mom's turned right. Craning her head up to the left, she saw dim, cloud-dampened daylight filtered through aging, threadbare curtains on the one small window in the stairway. Stepping up, she counted to make sure there were not thirteen steps. Memories of an old horror movie. The seventh and ninth steps creaked ominously. So what? No one could hear her steps. She took a few more and stepped onto the landing on the second floor. Down that hall, "the room," the door shut. All the doors were shut. That was why the view was always dim, unlit. No doors were open off the upstairs hall to let in light, and no window was in the hall itself. A set of empty built-in shelves lined the back wall where a window might have been.

Parker passed to the first door and pushed it open. Illumination went from dark to dim. She took two steps right and pushed on the opposite door. Streamers of southern light illuminated motes, disturbed by her passage, drifting in an empty room.

Finally! She could see color in the hall; the walls were pale yellow, the floor reddish. Well, they had once been painted a dark red; maybe in imitation of a mahogany or cherry stain. A few steps more would take her to the brink of every dream that had ended over the last several weeks. She stopped two steps short. Immobilized. She tested her new resolve with her right foot; it moved forward and she shifted her weight onto it. Left foot, shift. In two more steps she was there, the knob just out of reach.

Outside, a predator bird shrieked, and Parker jumped back a step. She took stock: heart rate up, breathing rapid, palms sweaty, tongue dry. There was a water bottle in the Jeep. Hah! Not a chance. She took the step, reached the knob. And stopped. The cut glass knob glittered with light leaking through the mechanism. But there had been a — a what, force field? Had the reactions in the dream been only a mental block, a willed representation of fear of the unknown? Mental what? This is the real world. She grasped the crystal knob and turned.

With a little push, the door swung freely into a large room spanning the front of the house. Two windows on each side wall and three across the front let in as much cloud-thinned light as its occupant would have needed to write, read, do after-school homework, or whatever pastimes were enjoyed in early twentieth-century America. As she looked out beyond the front window, a band of sunlight passed over the house, and she watched the yellow beam light up the landscape as it passed across the valley. It was comforting, somehow. This was a house built before rural America was electrified, and the room was a comfortable and enviable safe harbor with what was probably a wonderful view of the valley across the road when it wasn't raining or burnt to cinders.

She stepped into the room and felt the bottom drop out of her world.

4

Her first reaction to reach out resulted in a sharp pain in her left arm as she bumped an unseen edge that wasn't there a second ago. She had reflexively shut her eyes when her sense of balance abandoned her. When she hit the floor, she knew something was wrong. There were voices; why were there voices?

"SHYTE, she back!"

"Who?"

"Da wim dat wassere two week down."

"Crap me no?"

"Nah. Crap nah. She in da port layin' out on da flo."

The voices continued in jargon that Parker understood obliquely. Aware now that she was lying on a floor she began to move, tentative motions at first, to put her hands beneath her. She pushed up and got to her hands and knees. She opened her eyes again, hoping for a farmhouse floor. A polished stainless-steel floor had materialized where pine floorboards had been. Looking around, she felt as if she had woken up drunk after a sorority house all-nighter. Pain radiated from a hard impact above her right ear. A "why-am-I-here, what-the-hell-happened" sensation included an intense need to empty her gut. Her diaphragm heaved and heaved again. She spit, then, grimacing, ejected her morning coffee and bits of bagel. *GOD! I hate that smell.*

"O man, she pumped it," a voice to her left said.

"No probs, daas why iss stainless in dere." Then another voice, hollow with distance. "Get you on in dere, hep her up."

"Cha." Hands grabbed her arms, lifting. Another set of hands joined the effort. She let herself be lifted because her legs felt like stupid unresponsive children; they would not do her bidding. Her left knee and elbow complained of their impact with the steel floor. She shut her eyes against too bright illumination and the confusion of what they saw when open.

"Der, yo beder?"

Parker blinked hard, opened her eyes again. The young man standing in front of her wore a thin pullover shirt that looked like any tee shirt. Only its fabric was iridescent, beautiful, captivating. He was possibly mid to late twenties, pale skin tones. No sign of a tan, his features looked eastern European and — familiar? His hair, oh! His hair was blond, no yellow, too yellow in front but graded through brown to violet at the back of his head. The color job looked to be several weeks overdue for a touch up. Sandy blonde roots underlaid the color job.

Why am I worried about his hair color? This is not the room, not the room I stepped into. The uncomfortable echo of *déjà vu* was missing. This was not *déjà vu* ; she knew. She was positive she had been here before.

"Grieg?" a disembodied voice asked.

"Cha." Grieg turned his head toward a glass panel in a wall where she knew windows should have been. Had been, only moments before.

The one called Grieg called out again. "Chee member fine good."

"Cha," Parker heard herself say. "Grieg." She did remember, memory flooded in like sunlight through storm clouds. "Grieg, how could I forget so much?" Then she remembered a little more. "There are two more, right?"

He laughed a rich, deep, full-chested laugh. It subsided into a broad smile; long lashes fluttered. Did he have false eyelashes? She widened her eyes involuntarily as he peered with appraising eyes into her own.

When he spoke again, his speech was generic early 21st century. "Apparently, your injection did not take."

"Injection?" Despite the smiling face, Grieg could be dangerous. A bead of newly formed sweat ran from her temple past her ear and paused at her jaw line. She casually raised a shoulder to wick it up. She said, still groggy. "What kind of injection?"

"Yeah, the last time you stumbled in here. You were scared, manno. I think it would be scary. We had a brief convo, to reassure you that we weren't alien invaders. Noel give you some coffee that was a reasonable proximation of what you might find at one of your cafés, er maybe a store bucks?"

29

"Starbucks, they're called Starbucks. I don't know why." Parker licked dry lips that still tasted of bile, especially uncertain now if she should ask for a drink.

"Well, that star bucks," Grieg enunciated carefully, still working on the early 21st century idiom, "was laced with a sedation, uhm, sedative." He saw that she was sagging visibly. "Are you kay?"

She leaned back to put her weight on the counter that had not been there before. Before what? Before I stepped into that room, back in the farmhouse in the 21st century. "Yeah, Cha! I'm good. So, there was more than a sedative?"

Grieg looked at one of the vague outlines behind the glass panel. Was that glass? A nod of affirmation was obvious. Parker saw communication in his eyes to whoever was in the other room, He looked back at her and said, "Yes, we gave you a proper cocktail. It probably felt like you'd been bitten by a pretty good sized skeetoe."

"Mosquito," she enunciated.

"Cha, Mo skee toe." Grieg repeated. "That was a pretty powerful memory blocker. It was developed back in the 20th C. to keep patients who had to be awake during a proceeedure from remembering the ordeal."

"Really? What kinds of proceeedures?" She mimicked his incorrect enunciation.

"Not real clear on the research, procto exams, childbirth. Mebbe if more of your women membered how much childbirth hurt, there'd have been less of you."

"Wait, what? No." Parker shook her head in an attempt to change the channel. What the hell have I gotten into? Rubbing the heels of her thumbs into her eyes, she took a deep breath to calm her racing heart. The breath turned into a yawn. "Can I get a glass of water? My mouth smells like death and tastes worse." She put on her serious lawyer face. "Real actual water, with no special additives?"

Another nod of assent and Grieg turned to an instrument counter, pulled a plastic cup from a recess, filled it from a small thermos, and handed it to her. At first taste she winced.

"Sorra, is best we can do. Mebbe need to clean off the precipitators." He gave a casual shrug. "We don' get out much."

"Out where?" Her quick glance around the room fell on what she now knew to be an entrance hatch. More of an air lock than a door.

Grieg turned to see what she had focused on. "Mebbe you not 'membering as much as I tot."

"Maybe not — what's wrong with this water? Why can't you go out? Christ! I've got so many questions." She took another tentative sip, swished it around, and realizing she couldn't spit it out, swallowed. "Is there," she paused, a question forming. She could see other areas beyond the small space into a darkened doorway. Too much space for the small bedroom she'd entered. "Uh, can I sit down?"

"Cha, Cha, uhm. Yeah, come in here." Grieg led her from a small vestibule into the adjacent larger space; waved his hand at a wall and the ceiling lit up. She followed him, glancing up at the uniformly bright ceiling. The light illuminated a shadowless room with side entrances to other rooms. It was dominated by a small table that could seat four at the most. Countertops along both sides of the room were lined with unidentifiable instrumentation.

She blinked at the glare. She remembered the question. "Is there any reason for it to be so freakin' bright in here?"

The young man called Grieg waved a hand at something, and the room lights dimmed to a shade above twilight. He moved toward the far wall and tapped a console's key. A panel lock clicked, and a screen appeared. He turned toward her. "Come, look. You might as well see this. We'll have to try the sedative again anyway." He sat at a counter with no visible keyboard. A transparent blue sparkling energy field began to glow where a screen might be at her own workstation. He made an odd motion toward it, and the energy field became a display screen.

Unruffled by the obvious advances in computer displays, she came over and took a seat beside him. As he typed, or stroked at an invisible keypad, incredibly fast she thought, images flew past. He brought up a view of the Earth from high orbit. As the screen resolved, she realized it might be a live shot.

"Recognize this?"

"Of course, I'm not an idiot."

"This is your Earth, about 2020 or so." The point of view showed the northern hemisphere, with Southern California directly under the camera.

"What do you mean, 'my Earth'? It's our Earth, the Earth. Not YOUR Earth. It's an idiom thing."

"No, I mean it's your time, 2020." He tapped a few keys and brought up another planetary view. He guided a pointer without seeming to touch anything. His hands simply moved as if he were moving a grapefruit-sized ball, and the globe rotated until the point of view was the same as the prior image. "This is my Earth."

"So, why is that not our Earth or The Earth. I don't get it." She stared, comprehending but not wanting to believe, to accept what she was comprehending. "Are you saying I'm in a parallel universe or something?" She looked over her shoulder at the white on stainless room behind them. "This is not a farmhouse south of Bakersfield, in 2019. It's not the room I stepped into, what, fifteen minutes ago."

"Not fifteen."

"OK." She considered. Maybe she'd been unconscious on the floor for a few hours. "This morning?"

"No." He reached out a hand and placed a gloved hand on her forearm. The slightly latex smell was familiar at least. "Not this morning. That's the hard part. It's November 7, 2214."

She felt annoyed at the lie, then angry. Or worried, or frightened. That was it, wasn't it. The cold shiver began in her scalp and raced down her neck, spreading outward, raising the hair on her arms. Clenching her fists, her palms clammy; her mind wanted to shut down or wake up. The battle between confusion and comprehension had taken only moments, but the fear remained. Instinctively, protectively, she drew her hands into her lap. When a deep hatha breath, then a second, failed to resolve anything, she took another, sensing this time a metallic tinge to the highly processed air. She'd closed her eyes from yogic habit and opened them to the reality of a 23rd century interior and a digital scene from sci-fi movies.

2214! November 7th, 2214. She leaned toward the display and looked closer at the screen. She moved her free hand toward the screen as he had done. The view shrank away from her. Involuntarily, she pulled her hand back. In response, the view zoomed in close, very close. The viewpoint was only a few miles above ground level. She looked again. The entire valley, if the valley was the one outside that upstairs bedroom window, was covered in urban grid. The fields were gone. Gray and browns were the dominant colors. Green was confined to what looked like a stream bed and a few hollows in the mountain slopes. "What the…?"

He watched her getting used to the movement-sensitive controls. "You could get instantly better at that with haptic inserts, but there aren't likely to be any of those for you to use for forty or fifty years." He pointed to the display. "That's my Earth." He seemed about to say something more.

More memories from the previous visit began to form. "When is your Earth?" Then she turned to him. "What are haptics?"

~ ~ ~

The room lights came up. Parker heard footsteps behind her and whirled about. Two men wearing fairly ordinary white cotton T-shirts and something that could have passed for styled blue jeans came in from the other room. The smaller one had an olive complexion and curly black hair; the other was an Asian-Caucasian blend from first appearances, tall, thin, and bald as an egg. They were smiling. She wasn't.

The smaller, darker one spoke. "Ms. Parker, I wish I could say it was nice to see you again, but, you see, you are a bit of a problem for us."

"That's Parrish, I'm Parker Parrish." She hesitated only a moment. "How can I not be a problem for you? I'd be happy to leave. I swear, if I told anyone what's just happened, no one would believe me anyhow."

"Cha, er, yes. I'm sorry. Ms. Parrish." He extended an elbow out for an aborted attempt at an elbow bump, corrected himself, and somewhat awkwardly extended a hand. "I am Noel Antonides. I'm sure you've had quite a shock. We don't usually have that reaction, but then, we are expecting that little hiccup when we walk into the portal; you were taken, ah, unawares I think you say." He rubbed at a trimmed goatee. "So, you don't remember the entrance from your last visit?"

"No. Well, I remembered the farmhouse; I've been dreaming of it for weeks now. But my dreams never let me pass through that door." She turned and nodded toward the doorway. It was now a shimmering rectangle in gold tinged rainbow colors.

He crossed his arms, seeming to assess her. Then, "We get a little bit of the nausea still. But we have to go out in your time, occasionally for supplies." A grin flashed across his lips. "Or returning wayward visitors. Chu and I had to get you home."

"Grieg stayed here?"

"He rarely goin' out in da 21st, even to smell the fresh air. Chu and I, we do take the opportunity to step out in the 21st." His lips thinned and something passed across his eyes. He seemed to autocorrect and said, "It's in our job description."

"What's that? What's your job description? Mr. Anotid…"

"Antonides. It's Greek, or rather, from Crete. Ancient family name. You can call me Noel, and Parrish is — ?"

"I really don't know." She wondered about the odd introduction. "Why is that important?"

"Well," Antonides shrugged, "you 'Mericans blended the gene pool quite a bit in your day. It was an amazing period. We could probably look it up for you if you'd like."

Grieg sat beside her and offered a damp towel and a cool tumbler of water, "Here — this is simple water to clear your mouth." She hadn't noticed, but he must have wiped the floor clean too.

She took a sip and swished, swallowed. The second sip was better than the first. What was Antonides asking? Ancestry? "Don't bother with my family tree. I could research it if I cared much. I don't. I had always thought Parrish was French or English; it doesn't matter. I'm actually a lot more interested in how to get back to 2020. As curious as I was to get to that room." She paused briefly as the thought revised itself, *This room!* She shivered as another sudden chill ran down her spine. "I'd really like to go home now, get a glass of wine or three, and try to forget this happened. You've got to understand, no one will believe any part of this… this…" She waved her hand around at the room on their side of the portal.

"You're not interested? Your ancestors? Descendants? Want to see how your family tree branched after you? Any nuts fall to the ground?" She didn't like the following smirk. "We can see if your children are heroes or criminals."

"Why the interest in my family tree? If I want to look it up, I'll hire someone to go online and figure out how the Parrishes came to Florida from wherever the hell, probably Wisconsin, or Massachusetts, or some other frozen hell. I really don't care." *And it's not their business that I can't have children.* "What is this, some genealogy field trip from the distant future?"

The man she remembered as Chu turned to face her. He held a simple silver rectangle in his hand that looked at first glance like any of the several remotes required to operate her media center. He didn't make any mention of it; it was just held, almost menacingly, in plain sight. "Parker? May I call you Parker? We don' want to come off as hostile or o'erbearing." His complex accent was flavored by an unidentifiable third-world undertone with a heavy dose of the Caribbean islands; all delivered in a rich baritone. It took Parker by surprise. "We just fine ourselves in a predicament dat we didn' 'nitiate. You did."

"I ... I." Shut up. Stop talking, find a way out of this incredible screwup. They didn't kidnap me, I fell into 'their' future. "I'm sorry for intruding. If I can go now, I really don't need to know why you're here or what sort of quest you're on. If you set your, gateway or portal, or whatever you call it, and set it for when I showed up, preferably just after I arrived, I'll happily be on my way. Just let me leave?" Her inflection left the statement as a question.

"M'afraid we can no do that. Would violate too many of our pro-to-cols." Grieg looked quite sincere, his tone comforting and convincing at the same time. "We canno just let you leave. Despite your 'surances, you might reconsider and decide in a best-case scenario to just drop in for a visit, or in a worst-case scenario, to report our portal to some civic agency. You still called them 'the cops' I believe." He shook his head disapprovingly "That just would not do, you see?"

Suddenly tired, Parker leaned back on a counter. She remembered it might not actually be a counter but one of their invisible keyboards and turned to look at it more closely. Safe. She leaned back on it again; her rib cage seemed to deflate. She hadn't slept well Saturday night; she'd gotten up in the pre-dawn and headed north. How much sleep *did* I get?

Grieg stepped close and grabbed her arm; she was fading. The water! They drugged the water. Again! Her fading thought was self-incriminating. *You idiot.*

She fell onto him. "What? Have you done something to me?"

"No," Grieg said. His voice was calming, a comforting tone reminiscent of her father's calm but firm guidance. She focused on the sound; he was still talking. " —coming through the gate, as you called it, can be exhausting, disorienting for sure, and nauseating. I don't go through that door often, but never with more than a snack in me. I don't like to look at my food twice." He helped her to a stool that looked like it could have just been bought at any furniture store in Orange County. "You going to be OK? You're looking a little pale."

"Just water. Please, could I have another glass of water?" She emphasized, "Very plain water?"

Antonides was already on his way to the adjoining room. Chu asked, "Parker, what de highest level of schoolin you have?"

"I'm a juris doctorate." She simplified. "A lawyer." There was no response to this. "That's a PhD in Law."

"Tank you, I know dat. Dat's about five or six years after primary, er, your high school levels in 'Merica? Cha?" Chu seemed to be considering something.

She tried to explain. "It depends on if you are working full time and going to school at night. But normally, yes, five or six years. Four years of college with specialties in pre-law. Which is useless unless you want to file briefs or be a file clerk at a legal office, then two to three more for the doctorate."

"So, in your time, de twenty-firse, you'd be tot to be an educated woman?"

"Yes, I suppose so. Why?"

"Hones up, I'll have to do some checkin' wid some Bigs." Chu smiled; an eyebrow raised. "Dere's always been a chain of command, and dere still is. I have an idea I'd like to follow on up." He turned to the doorway of the other room, raising his voice. "Manno, war dat wat?"

"Come na later! Lax!" Noel called from out of sight. He emerged carrying a plastic tumbler with iced water and gave it to Parker with a faux bow and flourish. He stood, mocking a waiter's attentive air. "Madam, your pleasure?"

"Thank you." She took the tumbler and sniffed it carefully before taking a tiny sip.

Chu said, "Don warre, oop!" He paused to reframe his brain into the patois of the past. "Please excuse me. Don't worry, it really just a cold cup of water."

She took a deeper drink and enjoyed the cool wave as it passed down her gullet. She placed the edge of the tumbler against her forehead, feeling no refreshing coolness. She took a closer look at the tumbler. Its base was actually a movable dial. An indicator could traverse a scale that ranged from snowflake to sunshine. It was set to just above snowflake. She thumbed it closer to sunshine, and steam bubbles began to form.

Chu nodded. "Rather have tea?"

"No," she said. "Feeling a little better." She moved the dial back to its original snowflake chill setting.

Grieg asked, "So, you OK now?"

"Yes, thank you." And then, "I take it there has been some slippage in Standard English since my time." She emptied the tumbler and set it down. "What year did you say this is?"

~ ~ ~

Chu answered. "Dis is 2214. You from 194YBP." He too carefully spelled out the WYE BEE PEE. "That's years before present. To keep ourselves straight we have to keep track of both."

She looked at the three of them looking back at her. She was the anomaly. For all their strange fabrics and dialects, they could be understood at least. Antonides seemed to defer to Chu; Grieg was helpful but certainly not in command. Chu must be the one. She addressed him directly.

"You said you'd have to follow up on an idea? What's your idea?" She smiled through pursed lips at the absurdity of her next thought. "I have to get home to my cat and work up some arguments for a court case later in the week. But then tomorrow is, what did you say, 194 years in the past?" She stopped thinking. Her reality flooded in. "SHIT!" She folded her hands over her face and shuddered, trying to suppress the urge to cry.

Chu put a bone-thin hand on her shoulder in a comforting gesture. He said, "You be right, pleny soon. We have strick pro-to-cols against hurting anyone in your time—" He stopped in mid-sentence, leaving Parker unsure. Wary of another stronger dose of memory blocker, or some other solution.

Antonides picked up the line in a safer track and in more eloquent 21st speech. "We can't hurt anyone who might be an ancestor of anyone important in the time before our present. Not that we'd take out a slag on purpose, but we do have to be careful."

"Slag. That sounds like there must be some of us who are less important than others."

"Perceptive," Noel said. "There are millions of people who will make absolutely no difference on any of the events that shaped the course of the world in the past almost 200 years that would not be missed by history. It has always been so. But we can't always be sure who, even an apparent slag, might be the ancestor of someone who did become an important influence."

"You, Ms. Parrish," said Chu. "You da problem."

Noel, nibbling his knuckles, added, "You have proven to be resistant to the memory blocker we normally use to send back any of the very few people that stumble into this station by accident."

"I didn't plan to … I," She protested, stumbling over her tongue. With their emphasis on descendants, she did not want to bring up her own inabilities. "I was having these awful dreams about … about this house. I've been waking up more and more often with a recurring dream of having been in the hallway just outside the door, to your, your time gate, whatever you call it. And I couldn't seem to touch the door. I wake up in shivering sweats, exhausted and frightened usually. Lately it's been costing me rest I need. And I… I'm not sure I know how I got back here today."

Chu turned to Noel. Something seemed to pass between them in facial communication. A question asked, answered with barely a perceptible nod. Chu said, "Don' know about today. We put you back, lass time mebbe two weeks down."

Responding to her open-mouthed surprise, Noel said, "We have a car from the 21st. It's in that bit of a shed out back." He gestured back through the shimmering doorway.

Her eyes knit in confusion. What did that mean exactly? Would they not drug her again? Could she leave? But a nagging thought returned. How did she return to this spot? This exact way-out-of-the-way place in rarely visited California parkland. "You could let me go. I'd be very happy to go back. Forget I'd ever been here."

Chu asked, "How you get back here, dis time. You say you no 'member."

Parker continued, "I was out taking a Sunday drive, losing myself in the hills, taking random turns, and I ended up at this damn house again."

"Perhaps," Grieg offered. "Mebbe not so random a drive. What seemed to be this, then that, was in fact an eventual result of your predictable behavior."

"But I had planned to drive up the Pacific Coast Highway! That's miles from here."

Grieg mused, "That must have been beautiful; I read about the PCH."

She looked at Grieg. His crystalline blue eyes were staring directly at her, into her it seemed. He was not an alien—she blinked, surprised that she had even considered it—just a distant possible relative. He was a human. She asked him, "Why shouldn't I feel something ominous about that use of the past tense?" Then she realized. "Oh, it's not there anymore, is it?"

He shook his head in the negative. "Sorry. The 'big one' finally hit in 2062, and it was overdue by half a century. They never did decide if some of the quake-proof engineering helped or made it worse. Losses west of the fault were tremendous. Big buildings fell on little buildings. Ancient automatic shut-offs on the gas mains." His voice trailed off, apologetic.

"When did it? No, I don't think I want to know more about it."

"I 'gree," Chu said. "Come, lemme show you sometin. I wan your reaction before I go to big boss in Wellington wit your case." He pulled her gently to her feet and led her to the workstation that she had shared with Grieg. Sitting, he waved at the screen with some rapid movements and brought up the Earth view she'd seen earlier. He motioned for her to sit beside him. She did. Antonides and Grieg gravitated to positions behind them. When the image resolved, it was the slowly spinning sphere she had seen earlier. Looking at it from a full zoom, she noted that what she thought should be a Brazilian rain forest was brown. Clouds partially covered it, but it was not the deep green she usually saw.

On the African continent, the area she'd thought of as deep jungle was a lot farther south than it should have been. The sub-Sahara had grown. India was scarier; there was almost no green on the subcontinent. She wasn't sure, but she thought there were supposed to be jungle green areas in northern Australia. South and east of Australia, Tasmania, and New Zealand still showed green but no glaciers. She didn't wait for the slowly spinning sphere to pan the Pacific. She reached hesitantly toward the screen, then gestured for it to roll toward her. The North Pole showed a small patch of ice, not much larger than Iceland. Greenland was green at the fringes. Lines of snow-capped white accented the pattern of still glaciated valleys that had been smothered in snow since before humanity left Africa. It was supposed to be white! She waved down again, and the ball tilted over on its axis. The South Pole was still under ice. But there was a coastline and a green belt north of the ice line.

"We really screwed up, didn't we?" she said to no one, barely above a whisper.

Chu, in quiet tones, said, "Wasn' just you, your generation. History lesson say da fall been building since de nineteenth, but yours was the last generation that could have stopped it, or slowed it, and didn't. It kept accelerating."

"What's the population of the Earth now?"

There was silence for a while, and then Chu spoke. He had turned to face her. "Liddle more dan half a billion worldwide we tink."

She gaped open-mouthed, taking in the implications. "We'd gotten pretty good at counting, at estimating. What do you mean you think?"

"We don't mix as much anymore; dere is very little travel between livable areas. But dat's a whole udder story," Chu said. The two other time travelers returned to their tasks.

She asked the burning question, not wanting a yes. "There was a war?"

"Yes."

"Nuclear?"

"Yes, but twasn't total."

"Still…"

"Cha," he nodded, "pretty too bad. Sky link go down many places. Little wars all over. Bad guys take over Africa, Sout' America, go barbarian, some go cannibal." When he paused, she thought she'd detected emotion coloring his former calm. "For de first generation of survivors, it twasn' ver good to be a human being on planet Earth."

"Oh God!"

"I'm sorry you had to hear it from us. But, bleeding hell, no I'm not." Chu's voice grew agitated, and a bit less of the Jamaican lilt colored his speech. "Your generation wouldn't pay attention to de voices in de wilderness, or even de damned voices in de media. Plenny of people warned 'bout de atmosphere, 'bout the too hot ocean. You exceeded its capacity to hold carbon, metane, a host of others. You can'na tell it from dis," he motioned toward the globe that was still slowly spinning on the screen, the green rimmed southern continent pointed toward them. "Dere no more whales, elephants, large cats, most of the animals you all kept in zoos because de were interesting to look at. Dey are now video exhibits in a few cities wit zoos." He gave a disgusted grunt. "You made a lot of videos."

5

Parker felt ill. Her forehead bloomed beads of sweat. The room lost focus briefly. She closed her eyes, waited for her guts to recede from panic mode. A cold chill raised goosebumps in a racing wave down her arms. Had they drugged her again?

No. She forced her eyes to look back up at the screen. This is just some sick crap, to have to face up to the slowly-spinning orb displaying a gray to tan Siberian plain. So, the steppes had thawed and burned or been poisoned to desert wasteland. She heard voices talking to her through a tingling buzz, some in the local patois and some in recognizable 21st English. She forced a deep breath to press against her diaphragm to regain control over her flip-flopping stomach and blinked hard to suppress tears. Grieg was staring at her, concern etched into slight wrinkles between his brows.

"Grieg, isn't it? I seem to remember you from before. The last time. Your first name is Grieg, or Greg? Gregory?"

"Na mon. Na, but you close. Meh Grieg, G – R – I – E – G. Phonetic I s'pose to your English Greg. My Mom say she me name Grieg because Griegor sound too Russian. Aside from me Pop, she hated Russians. Me pop name Griegor, short for Griegoryev, for his pop. I'm from da Jamaica. Most of us, we survivors, are from islands. Noel," he pointed at Antonides, "he from Crete. Chu, we met in Jamaica; but from an American enclave in de Philippines, Chinapino 'Merican. Kind of a bastard if you axe me."

She looked up at Chu who was smiling and shaking his head, indicating no, no, no. He seemed to be taking the slur to his ancestry good-heartedly. She asked Chu, "How did you get a Jamaican accent in the Philippines?"

"Grandad and Ma got one of the last planes out of the Philippines, before the Second Death. Found another island to escape the plagues. Islands were the safest places to be." Despite the dour subject, Chu had a hint of a smile. "Dey went to de relatively safe island of Jamaica. Ten years later de died — the plagues caught up — I was a world class refugee at twelve wit dis ganja inflected accent." He smiled again, looking inward. "I stowed away on a trade ship bound for NZ where I met this Greco-commando programmer." He paused, placing a hand gently on Noel's shoulder. "Noel got me first into WuShu and then into the Time Service." He pointed to Grieg, "This sad reprobate's bounce was through St. Croix and Jamaica, then to NZ. Different ship. Now almos we all infected wit dis island jabba."

"Cha mon!" Grieg taunted.

Noel repeated, "Cha mon!"

The humor helped, but the problem persisted. Parker had invaded their space, their project. I should not be here. They won't want me to return. I should not remember being here, and I'm not a good candidate for the memory blocker.

Chu turned abruptly, spoke rapidly in their dialect, presumably to keep her from understanding, and went into the next room. He grabbed a breathing mask from a wall hook, entered a small enclosure that seemed only a little larger than an airliner's toilet. There was a flash of light as the translucent upper half of the enclosure flared white.

Parker couldn't help but notice the flash. "Where'd he go? Further future? Further past?"

"No, nothing that complicated," Grieg answered. Antonides had stepped over to a workstation and was palming through files on his screen. "Chu's gone outside to make a call. We have a transmitter that bounces to a remote on the coast and back to Wellington. Here is now. November 19, 2214, AD."

She regarded them — who was in charge? It was Chu who'd gone outside to make a call. She realized she had only partially followed Chu's journey to New Zealand. She asked, "Chu said he was from the Philippines? He met you there?"

Noel poorly mimicked the islander's accent. "Nah Mon! He down in Jamaica from boy chile. I find him creepin' off dat boat in Tauranga, little port city. He had mean street skills. Recruited at eleven, train to sixteen. He's still a pretty good fighter too." Antonides abruptly adopted a first position stance. "Nine years now in Time Service." He bowed, grinning with too many teeth, in an overtly obsequious dip.

43

She looked toward the booth that Chu had passed through. "Is that a portal too?"

"No!" he laughed. "It's our front door."

"But that flash of light, he didn't translate or something, go somewhere else on the planet?"

"For you, for now, top secret. I tell you, I have to kill you." Noel opened his arms palm out as if say, sorry, that is my only option.

"In that case, never mind." She took another sip of the water. There didn't seem to be any pharmaceuticals in it, so she took a deeper drink. "Listen, Noel. How can I get back home? This is fascinating, unbelievable really. But I have court in the morning, in my tomorrow morning, and a cat that will miss me, and people will miss me if I simply disappear. If my car is discovered out in the front yard of that house, there will likely be a lot more visitors to the gate. Sheriff's deputies, the cops. You're not going to want to get a bunch of scared cops falling through that gate. They tend to shoot at whatever scares them."

"Cha, we know. Saw the vids from the late 21st; it doesn't get any better."

She looked around at the room. She'd almost taken for granted that it should be exotic. According to these guys, she was almost 200 years into the future, and on a ruined planet. Something didn't quite fit. She asked Grieg, "If there are so few people left, if the planet is as bad as you're trying to tell me, how do you have all this?" She waved her hands around, indicating the surroundings.

"It's not ours; the ones before, they built this." Noel shrugged. "They had a lot of cool shyte gone down. But us? We can barely build a printed circuit board anymore. A world economy is a dream. Too many places are lawless horror stories."

"Oh" was all she could muster, her head tangled in confused images of every Mad Maxian apocalyptic scenario she'd seen.

Grieg had been talking over her confusion. "… and we had a war or two. Things went bad, very bad. But we didn' go backward and sit around a campfire."

"I thought time travel was science fiction. How do you…"

"Sure, and de queen Isabella, she tot de world was flat. Lindberg tot de speed of sound was a speed limit. Einstein, he tot de speed of light was a speed limit. People can generally only tink in terms of der own conception of physics. You said you were a lawyer. How much do you know about early 21st science?"

"I've got an undergrad in environmental biology?" Parker shrugged. "I don't know, I used to get a few popular magazines to try to keep up. I know how to read a periodic table, but I couldn't tell you how string theory ties the universe together."

"Well that's good 'cause dat one had problems too. They were missing some key treads." She looked blankly back at him, not reacting to the intended pun. "You do know they were still looking for missing little bits, god particles, dark matter, and all dat."

"Yes, in Switzerland I think, CERN?" *Cern?* That idea seemed to come from far away.

"Well, shyte manno. You really wanna get in it? You care?"

"I guess I just want to believe it. That this is a time capsule, or gate or something I won't be able to rationally explain tomorrow."

Grieg got up and returned to the first room she'd come into and motioned for her to follow. Antonides looked up as she went past and gave an indecipherable shake of his head. She followed Grieg back to his workstation. He called back the orbital picture of the Earth and rotated it around to the southern tip of Florida. She gasped; it was much smaller than she had expected. He handed her the small silver controller with four arrows and a central button that toggled up and down. "This will be easier until you get haptics and sync your brain waves to it. But that takes training."

Parker accepted the idea of mind-controlled computers as a logical jump in technology even as she looked at the little controller. The buttons were simple enough to understand; slide commands in four directions and a zoom button. She toggled the zoom button incorrectly at first, then got it right. She got to a zoomed-in view of what was left of the Florida peninsula.

First, the color was wrong, then the outline. There was very little green, and most of the state was missing south and west of Okeechobee. Tampa Bay was huge, the remains of St. Petersburg were an island. She scrolled the view up the coastline and could see that the sea had retaken the swamps in southern Georgia.

"Holy crap!"

A flash in the other chamber, their "front door," distracted her momentarily. She supposed it was Chu coming back in. Over the sound of high-pressure fans, Grieg continued.

"Ain' nuttin holy about dat crap." Grieg turned to look over her shoulder at the glowing image. "The Keys, M'ami, Pom Beach, Lauderdall? There's still some high risers poking out of the ocean like so many sticks, but they won' last too many more decades. Rust, storms. They're just bird roosts and so many tousand hectare of artificial reef."

"Didn't they try to do something?"

"Hope sprung eternal and all that. They built dikes and levees. They walled off the ocean." To her ears his pronunciation of ocean had three long syllables, o-cee-anne. He shrugged. "But it's all gone to hell anyway."

"The war?"

Chu entered, his eyes examining her again, but passed into the crew quarters without comment. Grieg picked up the narrative. "No, after Hurricane Sebastien 2075. Dat sucker went straight up the east coast of Florida and Georgia like a northbound train." He shrugged. "They always knew it was going to happen. Dey kept building anyway. Sebastian blew it all away. Then the tick virus come, the Amblyomma." He exhaled in apparent frustration. "Made your Covid look like a summer cold."

She wondered what an amblyomma was, then thought about the hurricane's coastal track. "They just had one of those a few years ago. Hurricane Mathew. They said it was the first on record to do that."

"Well," Grieg said, irony dripping, "It was only sixty years till de next one and it got people talking about Cat 6 as de new measure for hurricanes."

She tried to imagine that, tornado winds on the scale of a hurricane. Couldn't. Using the little remote, she guided the view towards the Florida panhandle where she'd grown up. The coastline was near the southern edge of what used to be Tallahassee. The city's urban scar ran from the Apalachicola River to the Suwanee and flattened a little on its northern boundary with the Georgia line. Scraps of green in the brown cityscape hinted at emergent radiation-tolerant vegetation. A long barrier island lay along what used to be a county line thirty miles inland. She couldn't bring herself to zoom in. "I don't think I want to know after all." She relaxed against the back rest, suddenly tired. Tired physically and tired of all the bad news. "Was there any good news? Did we get anything right?"

Grieg answered, "Medicine got better; dey found a systemic solvent for arterial plaque in de digestive juice of a tiny Madagascar tree frog. It cured heart disease pretty much." He stroked the thin hair on his chin that passed for a beard. "Lemme see. Dey got really good at bone an organ replacement, and gene terapy stopped or cured a lot of cancers. Life got longer, but still not a lot better. Who needs a telly on da wrist anyway? In about mid-21st there was a backlash against the use of chip implants for keeping tabs of credit and that movement died. It was too hard for honest people to cheat. And it was too easy for wives to keep track of husbands and vice versa."

She had turned in her seat to avoid the screen and faced Grieg. His face read compassion for her sense of loss. With lips pressed almost to white, he shook his head in what appeared to be disapproval of the times. "The Third World stayed hungry."

She thought, "So much for human compassion." She felt a ball of gorge trying to rise and swallowed to keep it down. A bitter taste at the back of her tongue warned of an incipient urge to vomit. A bead of sweat rolled down her temple.

"Bless 'em, dey tried. The seed companies kept making super crops. And ma nature, she kept making super bugs. Then those crops failed." Grieg sounded apologetic for sharing the bad news. "Hey, Miz Parrish. Don' get me wrong. It was a very slow gradual ting for a while, like it always is. You creep up on a limiting factor and find a way to extend what you're doing. They built more solar, but kep burning gas in cars, until finally, juicing a battery got cheaper. When the crunch for petrochems came, they started digging up landfills for plastics. It didn't get really bad until the Arabs and Iraqi oils finally started to run out."

She felt her head nod and correct. Chu re-entered the little common area; he was looking at her with a studied expression. She took a deeper breath, stiffening her ribs to fight a heavier sense of gravity. Her eyelids batted against her will to remain wide awake. "I …want… to…" Head drooping, she was only able to form the thought since her lips refused to make the sound: *"Drugged. Dammit!"*

Parker felt herself slumping now, sagging in her own bones. Her breathing was becoming labored. She struggled to purposefully raise her head to look at Grieg's apologetic face. "There *was* something in the water wasn' th…?"

Her eyelids fluttered heavily and closed, just after she saw him shrug an "I'm sorry."

~ ~ ~

6:30 a.m. In a state of groggy confusion, Parker awoke next to certain that her experience in the front bedroom in an old house south of Bakersfield was surreal, as real as a dream can get without stubbing toes to prove it. Her current position, unhappily awake in the predawn twilight in her own bedroom, led her to an inescapable conclusion: she was becoming obsessed with a recurring dream. This last one was different. It had the detail and persistence of memory. It felt SO real! It had gone a lot further than the previous mental block at the hallway door; that had been easy, extremely ordinary. But three different characters inhabiting a room in the future? She'd given them names and weird ones too. Grieg and Chu? Noel?

God! She felt around beside her and found Jazzman curled on the other pillow. She pulled his limp body to her and stroked his neck. After a snort and shake of his head, he settled down and resumed purring. *At least one of us is content.* Padding into her bathroom, she ran wet fingers though her hair, washed her face and took a critical look at herself in the mirror. She was vaguely aware of a bruise on her elbow, and lifted it to the mirror to see that no sign on an injury remained.

Aging had been a forgiving process for many years. With minor enhancements like foundation and eye shadow, and occasionally some blush, she was able to stay at the top of her game, date most of the men she really wanted to date. And wow! That run through the desert with Carl!

Carl had been fun and thoughtfully attentive. He listened and inquired about her cases even though he had almost no exposure to the law. His own life and family were complex enough that he had numerous levels of understanding of family issues. But still, he was a techno-geek and not even thirty-five. She was comfortably into her forties. Ah, well, he'd been fun. But there had been that opening for a "later."

Over coffee, a half-bagel with jam, she noted the slowly brightening sky turning salmon pink to lavender and finally to the light blues of daylight. A light plop from the front of the house signaled the paper's arrival in the yard. In slippers and a robe, she ventured out to the street, picked up the paper and looked up and down the still-empty roadway. The curbside was empty; *where's my Jeep? Who's stolen the damned thing?* Fuming, she stomped toward the front door. What the—? It was on the far side of the driveway, a few feet from the garage door that hid all the things she'd never given away after ... after Douglas died.

She rarely parked in the driveway. Backing out of the blind entrance onto the treacherous curving cliff-side drive was unsafe, so she had always parked in the narrow delivery lane between her privacy wall and the street. Parker thought back to Saturday night. Coming home from the overlook when she'd dropped off Carl, she'd had a slight residual buzz from a shared bottle of wine. Perhaps, just maybe, she'd parked in the driveway to save a few steps to the door. *No, Christ! I had been driving the Jag.* It was in its usual spot.

She'd dropped the paper in confusion. In a groggy stupor, she picked it up, returned to the house, padded into the kitchen, and laid the paper out. Her second cup of coffee was interrupted by Jazzman's cries of hunger She fed him and settled into a chair overlooking the Pacific cliffs. She watched as a fog bank farther north moved in from the sea, obscured the houses up the cliff, and promised to steal the morning sky.

An intermittent buzzing interrupted her scan of the news. When the short buzzing burst repeated, she realized her phone was on silent. With most of her notification settings off, who would be trying to reach her? There was scant communication with what was left of family back in Florida, so what? She set down the article on corruption in the school board purchasing department and reached for her purse.

Reaching into its depths, she found the edges of her slender iPhone, but it seemed to be layered under keys, her mini travel bag of necessary cosmetics, and accumulated tissues and other detritus. Frustrated at her still-foggy mind and silently cursing the evil twin of red wine, her hangover, she emptied half of the purse's contents out on the kitchen counter. As she picked up the phone, it gave its third buzzing announcement, and she saw the green window displaying Carl's message: [Thnx for gr8 drive dnr and :) Up for a drive up the 5 anytime].

Grinning at the memory, she picked up the refilled coffee cup, let the warm vapor caress her nose, and reminisced. Yeah, it had been fun. He was fit, trim, and young. Too many of her partners were five to ten years her senior and were losing their masculine edge. *Wow, so now I'm a cougar? Am I going to start preying on younger men to get my needs met?* She tapped out her response: [;-) and thank U mister]. As she tapped send, she noticed the time stamp at the top of the display. 7:45!

Christ, it's Monday! She set the phone down and rubbed her temples. But the paper was fat and heavy. Her eyes ran to the top line of the paper. Sunday? But I read the Sunday paper yester—. *Damn, maybe I left the paper out there yesterday.* The phone's message had faded, and the screen gone black, so she tapped it on. She pulled up the messenger app again and this time texted: [Meg, feeling horrible, fluish :(Staying in today, hope to feel better by tomorrow. Put Stanhouser in my Dropbox. Will work from home. Sorry] She dropped the phone back into the purse and was about to scrape its contents off the counter to dump back in the purse when clarity began to take hold.

Still reeling from residual confusion of whether it was Sunday or Monday, she noticed a small silver shape in her purse next to her phone. With eyes widening in growing recognition, she dumped out the remainder of the purse's contents. A thin silver, black-button controller fell innocently among the upended contents. She picked the controller up, stared at it, and screamed.

6

At her scream, Jazzman darted from the kitchen. Parker dropped the silver remote back among her purse's strewn contents and stared at it in disbelief. Its confirmation of her remembered dream as a real-world experience left her speechless and gasping in short, spastic breaths. She sat, suddenly very alone, dumbstruck, mouth agape. It was exactly as she remembered the little controller from her mapping session with Grieg. She tapped the small button at the bottom of her phone looking for the time and date stamp. She nearly dropped the phone. Sunday? Sunday! How could that be? *I went driving yesterday, Saturday, with Carl. So when did I go to the house? When did I go for a Sunday drive and go to the farmhouse? How the hell did this little technical marvel materialize out of a dream and appear in my purse?* She ran back out to her driveway. The Jeep was still in the wrong place.

Parker cupped her face in her hands and struggled to reconstruct what might not be a dream. That last part of the adventure was a little fuzzy. The day she thought of as yesterday, also a Sunday, came back to her in disconnected scenes. She remembered, finally; the last scene she could recall was falling asleep in the middle of a conversation with Grieg. No, not falling asleep: passing out. *I was drugged after all!* She shouted at the empty street and no one in particular, "The sons-a-bitches!"

She considered what other confirmation she could draw on. The Jeep! She slid the key into the ignition and gave it a turn to the first click. She didn't need it to start; she just needed to glance at the gas gauge. There was less than an eighth of a tank. Bingo! She would never have left herself with so little gas in reserve. She normally gassed up as the needle slipped below half a tank; blocked freeways were too common, and she never wanted to run out of gas again in the middle lane. She got out of the Jeep and, walking shakily around it, leaned up against its rear-mounted spare tire.

Carl! Forgetting to close the door to the Jeep, she hurried back into the house and sent off a quick text. [Something I HAVE to show you, come as soon today as convenient]. She sagged against the door of the Jeep in confusion, staring vacantly at her phone. Then laughed at herself and followed the text up with her address.

The return message was almost immediate. With a bright chirp the phone announced: [On the way].

She tried to remember if she had ever actually given him her address; there had been a small quantity of adult beverage consumed, and maybe all the details weren't too sharp. Then again, he was extremely tech savvy. That was *one* of the reasons she had called him about the silver gadget. She thumbed on her phone again to double check its date/time display. Yes, it really was still Sunday. *You can't fool a phone; the signal comes from the tower.* She was back in her house shortly before she would have arrived at the overlook on Pyramid Lake on the "other" Sunday. Had they put her back to bed, before she'd gotten up? When was that? Does it matter, does "when" matter when dealing with, who? Time bandits, time lords, jumpers from the future? *Too much to think about without coffee.* She headed for the kitchen and her coffee pot.

Calm down, you're freaking. Her old shrink used to tell her: distract, don't react. Use your ADD to your advantage when you start to go manic. Parker didn't mind being borderline ADD; she got a lot done. *Ritual, that's good too.* She turned to her cabinets and started the process of getting out the dark roast, her chicory, and blended a new full batch of her mix. After all, company was coming. The paper, still Sunday, announced Saturday's news. Its bulk reassured her that advertising was alive and well in Southern California. Headlines in the local section warned of reduced but still high fire danger in the canyons despite yesterday's rain; the business section projected spikes in energy costs. One analyst was quoted as saying that the higher cost of energy made the cost of distributing energy more expensive and so on. *Duh!* she thought and wondered idly how Trevor Noah might make hay of such insanity. As the coffee maker began to burble and gasp, she heard a familiar car drive up. *Damn, I'm so out of it I left the front door open.* She ran for the bedroom to replace her robe with actual clothes.

When she heard the doorbell, she yelled out, "Come on in, I'm in the kitchen! Coffee's on."

~ ~ ~

She had dressed to impress in slim cut but not skin-tight jeans and a plain white blouse, three buttons unbuttoned. A single small gold chain hung just below her collar bone. She saw he was dressed for comfort in cargo pants and a logo-less pocket tee. Thankfully, there were no pens clipped in the pocket to definitively confirm his 'geekiness'. "Good morning, Carl. At least, I think it is."

"And good morning back, and thank you, Parker."

"For..."

"The desert run, dinner," his eyebrows lifted, "afterwards, and your text this morning." After a short pause, "Is this convenient?"

"Is what convenient?" *God! Why do I sound so stupid? He was good, but I'm not going to jump his bones.*

"Your text. You said come as soon today as it's convenient. Is this convenient?"

Fool! She recovered quickly. "Yes, and thanks. I seem to be on the verge of becoming a mad woman, but I have one link to sanity. A real physical link that you might be able to help out with." She turned and found the little remote-control device and handed it to him.

"Looks pretty simple. It's a remote, for what?"

"That story's a lot longer than you'd think and most of the reason I think I'm going crazy."

He turned it over a few times, pressing the five flush mount buttons in turn. Peering at them closer, he realized that they were not actual buttons, but pressure points. He provided an ongoing narrative as he examined the device. He'd seen some research that would replace many of the button board controls on a typical remote with a solid-state, pressure-sensitive switch. The trick was to make it sensitive to a finger's warm pressure and not any surface that might be laid down on it accidentally. The solution came from the gorilla glass used on cell phone and tablet surfaces.

At one end, there was a narrow slot that looked like the pry point for taking it apart. "Do you have a very small screwdriver?"

"Just a second." She turned, pulled her junk drawer open, and came up with an eyeglass repair kit. She handed him the tiny flat head screwdriver.

"Perfect."

"Sorry, I'm being rude. Can I get you some coffee? It's fresh about three minutes ago, and I really need another cup."

"Are you OK? Vin rouge hangover?"

"Yes, er, no. I mean, it kind of depends?"

"On…?"

"On whether I'm crazy or not." She pointedly looked down at the little device. She realized her hands were shaking and, needing to get them under control, grabbed at her empty coffee cup.

He held up the controller. "What did it do, what's it control?"

She turned to get down another oversized mug. As she set hers down, it shattered. "Dammit!"

"Easy, easy! Here, let me help with that." He was up and moving while she stood in semi-shock as bits of ceramic spun to a stop or careened over the edge of the marble countertop. "Just don't move, your bare feet won't appreciate how sharp these shards can be." Parker stood still but was overly tense. She was almost vibrating with tension and probably didn't need more coffee. Carl dampened a dish towel and began to wipe at the fine dust and bits that dusted the floor near her feet.

She looked down at him as he worked. A tiny thinning spot at the back of his skull suggested the preliminary stages of male pattern baldness. Then she noticed that he was looking at her legs and smiling. He carefully wiped the dish towel under her toes. He said, "We really don't want to hurt these tootsies, do we?"

She reached down and gently touched the black curls where the hair on the back of his head twirled up in half a dozen little duck tails. As he stood, he held her palm to the back of his neck so that, standing, she was in a half-hug. He leaned over and gave her a polite peck on the lips.

She took a large breath and released. "Thank you, Carl. I was on the edge of losing it earlier and I really can't afford to not be on top of this. It's way, way too strange." She started to go on but decided to let him have a look. She could brush everything that happened yesterday into dream-state baggage except for the nagging details of her Jeep's lost gasoline and this damned little remote.

"Well, it's exotic — state of the art, but not that strange."

"Can you crack it open? See if perhaps it's as ordinary as it seems?"

Carl continued wiping down the counter and picked up the largest pieces of the broken mug. "Coffee for two?"

"Yes, yes. I'm sorry. See? I really am a mess this morning." She reached up for the cupboard and got him a mug but didn't replace the broken one. He went to the other side of the counter and sat back down to his task; she filled his mug, set out sweetener in several different colors and a pint carton of half & half. "Suit to your tastes."

He tasted his coffee and smiled appreciatively. "It's good black. What's that flavor on top of the coffee?"

"Chicory — I order it in from New Orleans."

"I like it. I could get hooked!" He set the mug down and picked up the remote and the little screwdriver. With a click, the back of the unit separated, and they both peered over their respective sides of the counter to look inside. He pulled the back off and tried to understand what he saw. Almost nothing. A thin foil packet seemed to be pressed onto two contacts. It was clearly removable. He turned the top over and compared it to the nearly blank surface on the inside.

"This is really cool, but… there doesn't seem to be anything I'd call a battery. And this little packet of foil is the guts of the transmitter, so when it's together the leads go to this." He pointed the tiny screwdriver at a layer of black lining the backing plate and traced how the 'battery,' if that's what it might be, contacted the circuitry packet when the two pieces closed. He set it down and looked at some of the script molded into the inner surface of the backing piece. Three lines of script ran approximately the same length in English, Cyrillic, and German; two more lines of characters in, he supposed, Japanese and Chinese ran parallel to the western fonts.

"What's it say?"

He looked up at her, blankly. "Not much. Omega Corp. There's a number, and 'Made in Republic of Hawaii.' Republic of Hawaii?"

"Great!" Parker was clearly happy to hear this. "I'm not crazy!"

"Well, maybe you passed it on. Why don't I think it's crazy to contemplate the Republic of Hawaii?"

"Because in about 150 or so years, islands will be the safest places to live."

His blank stare deepened as his mouth slowly gaped open. He turned his head slightly, confusion squinting his eyes.

"Remember Friday morning? The dream about the farmhouse?"

~ ~ ~

Over the next half hour, she tried to explain what had transpired on Sunday, her other Sunday. For about that long afterwards, Parker tried to make Carl understand that it was all a perfectly good explanation for her to have, in her purse, an article of advanced technological manufacture.

She was becoming more frustrated with Carl's early and secondary resistance to her story. She said. "Listen, I have an idea. We can test at least a tiny portion of the function of that remote without having to point it at anything."

"What, without a 3D Earth viewer, what would you do with it?"

"Demonstrate that the black lining is a battery. Electricity should still work as we would expect it to, right?"

"I don't have any test equipment with me."

"Just hold on." She went to the garage, rummaged for a minute or two and returned holding an orange voltage multimeter.

"So, you're a little handier around the house than the average attorney?"

"Only a little. When my husband died, I learned to use some of his collection of arcane tools to avoid having to pay a hundred-dollar service call just to push a circuit breaker." She picked up the front piece and pulled gently on the foil packet. It popped off its two contacts, and she laid it on the counter. She added conversationally, "I can even replace a circuit breaker."

He gave a non-committal "hmm" as she untangled the two leads and removed the insulating tips. She dialed up the lowest voltage range possible on the meter, 200mVDC, and pressed the two wire contacts to the two tabs. The needle pegged out. She nodded to Carl who was holding the small instrument. He moved the switch to the next setting, 2VDC. Again, the needle pegged out.

Carl whistled. One more should do it. At the next setting, 20VDC, the dial settled on just under 12. In a state of disbelief, he wet a fingertip and touched the two contacts. His yelp almost drowned out the sizzle and spark. A slight odor of ozone and burnt flesh fanned out as he waved his hand to alleviate the unexpected pain. Parker produced an ice cube from the refrigerator door and a paper towel. She looked at his raised eyebrows. "So, do you know anyone who can pack that many amps into such a tiny package with that much zip to it?"

"No, and hell no! I'm beginning to think your crazy is rubbing off. Soon I'm going to roll over and my clock is going to say 6 a.m." He sucked on his still-hurting fingertip. "I'll turn on my coffee pot and wait for your text message and we can start this Sunday all over again."

"That's not funny! This Sunday has started once too many times already."

"Sorry, I was going for humor."

She leaned into the curved back of her bar stool. "I need you to try very hard to believe what I've been trying to tell you. I know without doubt that I found the house in my dream. That the upstairs front bedroom is a portal to some kind of transfer station in the future, and that they put me back here as if nothing ever happened."

"I suppose if the cops dusted your Jeep for prints, it would come up clean."

"Why bother? The owners of those fingerprints won't be born till about 2170 or 80." She sighed in exasperation. "They probably wiped it anyway."

"What time yesterday, I mean, the other Sunday, did you find the house? About when did you get to that room?"

"I don't know, I was driving randomly. Probably about 10:30 or 11:00. I guess I was inside for several hours." She looked down at her clasped hands. "I don't know how I got here, or when."

He looked at his watch. "Where'd you say this place was?"

"I don't know the exact road name. Remember I got there from over the mountain on park trails. But it's on the valley side, southeast of Bakersfield, I think. I pretty much found it by accident, and I did not leave there under my own steam."

"Right." He reached across the counter and took her hands in his. "Today I'll drive. Come on, you can scan for it on Google Earth as we go."

They took his Porsche and headed up through the hills toward Interstate 5. Traffic was light even for a Sunday, and the fog bank had not yet begun to climb over the coastal range as it had "yesterday." On her phone, she zoomed into a view of where she thought she had been, entered the coordinates into his GPS, and waited for the blue line solution. "Head for the 5 northbound. Looks like we could get there in two hours, if we decided to go the speed limit."

7

The rain caught them as they were headed up the interstate. It took them a little longer than they had expected to get to the house. The gate into the hills at the Pyramid Lake overlook was locked, and prying the chain bolts out of the timber frame with a tire iron delayed them. Parker puzzled over this because she'd found the lock dangling from the chain and left it that way when she drove through. The mountain road rose above the clouds before they hit the crest, and Southern California's central valley spread out beyond. To the east, a Mondrian landscape in shades of tan squared off by right-angular roads was punctuated by cookie-cutter circles rendered bright green by massive center-pivot irrigation machines.

She guided him by memory, taking the same formerly arbitrary turns she now clearly recalled. She had to slow him as they neared a seldom used turn off. The homemade sign said 'Private Property, Trespassers Will Be Shot' over the image of a revolver.

"You took *this* road?" He was incredulous.

"Yesterday it was overgrown by scrub, if it was even there."

In less time than she would have thought, they were at the house. As it came into view, Parker craned her neck to see if she could see her Jeep parked along the road. In the Yester-today, as they had begun to call it, she had been there by now. The cloud deck moving in from the Pacific passed overhead and the fire-darkened hillside gave off a charred smell as light rain fell."

They sat, motor off, and discussed the possibilities. Parker had been there, or here, yesterday. Yesterday's Parker was not here now, because she was here looking at the house, but it was Sunday, at about the time she had been there, in the house. Yes, the drizzle had started at about this same time. Did her removal by the time travelers erase what she knew as recent history? Or, she had not been there yester-today and was going a little bit crazy and it was infectious because Carl was still convinced the 'artifact' wasn't from here and now. Or, they were being punked and a television crew was going to show up soon and set everything straight. Last but not least, should they expect to see yester-today's Parker Parrish show up in the Jeep any minute now?

What about time travel paradoxes? Did one reality erase another? Was it impossible to violate two time lines at once? Nothing made sense. Had all the sci-fi writers gotten it wrong?

The rain shower ended, and a band of sunlight passed over the scene. She watched the band of sunlight travel downslope and then out across the valley floor. She gasped at the memory. "This time yesterday I was up in the house. I had just opened the door when that spot of sunshine passed through the cloud deck." After its passing, the scorched hillside regained its bleak, foreboding character. The house's unburnt yard was, if uninviting, at least evidence that someone had saved it from the recent wildfire. "Do you suppose those guys from the future had anything to do with this house not burning?"

Carl looked more closely at the house and its unscarred yard and garage. "Maybe. But maybe that's not really important. Maybe a fire response crew was parked here, and they just kept the area around their trucks from burning."

"I was just thinking that if they had wanted to preserve their portal, or gate, they'd have had to intervene with the fire service somehow."

"OK, but I think we're just avoiding the obvious."

"What's obvious about this place?"

She saw a worried frown dancing in his eyebrows. "At this time yesterday, *they* were here. Whether they erased your presence or not. It's also obvious that you aren't there. The yester-today they erased may have never happened. Yesterday you were here in your Jeep, it was parked, where?"

"Just ahead, about fifty feet closer. But —"

The absurdity of the what she knew and could prove and what she saw and couldn't believe hit her hard. Her reaction was laughter; it came out as a near silent shuddering pant and escalated to a breathless wheeze. She got control of her diaphragm finally, but not the absurdity. "What happens if we go there, and I show up while —," she stopped and giggled again. "Sorry, that's just not right. I can just picture it. "Sorry Ms. Parrish, you'll have to leave now. You are expected to drop by in few minutes, and you'll have to leave so you can get here." She found herself crying from laughing and stopped abruptly. She didn't want to sound as hysterical as she was beginning to feel. Not enough sleep and one too many iterations of the same day were getting to her.

When her breathing slowed, concern colored Carl's eventual response. "Parker, did you happen to see yourself yesterday?"

She frowned and faced him. "No, but that was the first time around."

She watched his profile as he studied the house, absorbing the details in the scene, turning back to her, then back to the humble farmhouse. He said, "Do you remember anything from your first trip to the house? When that was? How long ago that was? What happened then?"

"Well, no. They didn't say how much earlier the first visit was. Only that I'd been there. I guess I vaguely remembered the configuration of the gate room, or portal, I forget what they call it, but whatever they're using for a memory blocker isn't working a hundred percent on me."

They sat and watched another thirty minutes. She was intensely aware of his cologne but determined not to think of it, or him, or their afternoon at the cheap motel. The light rain continued, off and on. Finally, he spoke. "I think we're safe." He tapped her wrist. "Your Jeep hasn't shown up a second time. I think we're good. Let's go."

~ ~ ~

The stairway squeaked in all the right places. Down the hall, a door to the side bedroom was still cracked open from her remembered visit, letting a thin band of light into the hallway. "At least," she whispered, "it's not as dark in here." She reached out and pushed the side door further open, brightening the hallway several shades.

"That's the door at the end of the hall?"

"Yes, it's all here. All we need is some creepy violin music."

He chuckled quietly, more through his nose than mouth. "Keep 'em coming. My knees aren't shaking, but my reverse gear is ready."

Now she chuckled in a higher pitch, edging on nervous, and took a full pranayama breath to calm herself. "Are you ready?" He nodded. She turned to him. "Ready?"

"Yeah, yeah. Let's do it."

She opened the door to the front corner bedroom and pushed it open. As before, it swung freely providing a good view of an empty room. She walked in slowly this time, feeling ahead of her with outstretched hand. She felt a subtle sense of backpressure about two inches from the door. She remembered that from earlier, but she'd entered unawares, and the portal had grabbed her before she'd fully understood what was going on. If she pushed slowly, there was a palpable energy field pressing back.

"Come!" Parker said and stepped through.

~ ~ ~

Parker paused this time as she stepped through and reached out to brace herself on the table. Carl almost stepped on her heels. "Ow!" She tried to step forward and sideways as Carl came through. "Sorry, I should have gotten out of the way."

Chu was standing facing them. He waved at a small wall panel and turned off a softly pinging alarm. "Buddha's britches. Why do you persist, Ms. Parrish?"

Parker tried her most winning smile and a little Southern lilt. "Why, Mr. Chu, ahh just can't keep myself away from here." She gave an affected wave of her hand at the limited space. "Everthang is just *so* interesting." Dropping the accent of her Southern heritage, she reached into her jeans pocket. "And besides, I have something you might need." She extended her hand, holding out the little remote-control device.

Chu flinched at first, uncertain. His eyes widened with recognition. "Shyte sideways! How did you—"

"I don't know. I really don't have any idea." The recent scene of her re-awakening on her second Sunday came back. She opened her mouth to try to speak, shut it.

"So you back to return it."

"No, I brought Carl to prove to myself that I was not crazy."

Chu smirked at their nervous clumsiness. "Let me guess, you two are a pair?"

She shared a look with Carl. He answered, "Recently." Then, "Yes."

Chu shook his head, half-amused and half-confused. "Have a seat. Can I get you something to drink?"

"No, I'm not drinking your water again. Last time—last two times—you put a sleeper drug in it and sent me back."

"Dat was Grieg's doing. It wasn't me. I tot we had established dat memory suppressor didn' work on you."

Carl asked, "Is that a matter of policy? Accidental visitors get summarily shipped back?"

"Slags," Parker said, aside to Carl.

Chu said, "Cha, we have to take some precautions. Our pro-to-cols call for some extreme caution unless we identify the intruder as someone who wone affect the future."

"A slag," Parker said. "I forget who, either Noel or Grieg, said anyone who has no effect on history is a slag."

"Slag?" Carl said, testing the word. The dross that floats to the top of an alloy melt?"

"Or useless waste," Chu answered.

"That's pretty obnoxious if you ask me," Carl shot back.

"Nobody did. So, what about chink, gook, and I'm mebbe guessing, wetback?" Chu matched his stare, unflinching.

Carl stiffened, his face gray, eyelids narrowed.

Chu looked at Parker with an eyebrow raised. "Anybody ask him?"

She shook her head. "Carl, sweetie, I think he's making a point."

Chu explained it, staring hard, unyielding into Carl's stare. "Don' thick your skin, manno. Troughout history, all classes of people gave other classes of people demeaning names. Not so much for de humiliatin' or denigratin' effect, but to allow dem to do tings to dose other classes without tinking 'bout dey humanity." He looked to Parker, his expression softening. "Dat about get it?"

Thin-lipped in appraisal, she nodded. "Yes. Carl, we might or might not have any effect in the long historical run; therefore, we do not have the protection of their protocol as time lords or time bandits or whatever the hell it is they are." She took a calming breath. "Carl, in spite of their 'return to sender' routine yesterday, these guys have been polite, sweet almost. Grieg, where is Grieg? He was my tour guide yester-today, and Noel."

"Day out, down in the valley." Chu nodded toward the airlock-protected back door. Question marks flashed across his eyes. "Yester-today?"

"Which year are they out in?" Parker glanced up at the double-display calendar, its tenths of seconds flashing by.

He followed her glance and read the date. "Yours, March 20, 2019. What's yester-today?"

"Whoever took me back, took me back to before I left the house on Sunday March 20, which happens to be today. But they weren't totally clever, they didn't gas the Jeep."

Chu looked thoughtful for half a beat, then shrugged. "Best laid plans, and all dat. Robert Burns, right?"

She didn't answer. Instead, she asked, "Doesn't it get confusing? All the coming and going, the whens, and thens, and whenevers?"

Chu looked at her like she was a simpleton. Then turned to point to a display on the wall beside them. The stacked three-centimeter numerals read:

2020:06:20:10:42:16
2214:11:05:11:02:23

"Paying attention is all it takes," Chu said with a little condescension in his voice.

Parker bristled. "Hey, we might be from two centuries ago, but we aren't stupid. I'm one of the best goddamned lawyers in the OC, and Carl is one of the brightest tech minds you'll meet, at least in our time." She looked to Carl for confirmation, got a nod of embarrassed agreement.

"Easy, Parker." Carl tried to lay a calming hand on her shoulder. She pushed it away. He pointed at the display, its seconds rolling continuously. "So, it's November on your side? Are we in November 2214?"

"Yes, it's cool, in the evenings. Thirty degrees, still pretty damn hot daytime."

"Really? That's a hell of a daily swing?"

Chu cocked his head, then understood. "That's thirty Celsius."

Carl's face wrinkled in a frown; his eyes closed in calculation. "Ninety degrees is a cool November evening?" He scratched at the back of his neck. "Christ, that's –"

Chu shrugged. "Tings are different."

~ ~ ~

The three stood uncomfortably as the next few seconds of silence began to stretch into hard realization for the two visitors. A small white oblong over the back door blinked red. Chu waved a hand at a wall control panel, and the oval turned green. They could see through a translucent half panel that someone was entering what Parker had learned was a scrub chamber. Chu called out, "Dat lady come yesaday. She back!"

The three heard a string of expletives, or what sounded like expletives. But when Noel entered, his face was wrinkled in a hard frown. He shot a hard glance at Chu, forced a thin smile, and stepped up to greet the two arrivals.

"So, Ms. Parrish. So nice to see you, yah?" He bent at the waist in greeting, more in the European style than the Oriental.

Parker held Noel's gaze, extended her left arm toward Carl. "Noel right, Anthoneades?"

Noel grimaced at the attempt. "Noel Anton-I-des." The grimace turned to a neutral smile. He shook his head in apparent disapproval. To Chu he asked, "Troubles, manno?"

"Nah mon. Day be cool."

Parker gestured toward Noel. "Carl, meet Noel Antonides." She turned back to Chu.

Parker asked, "You're Chu, right? Chu what? What's your whole name?" She almost wanted to ask "who's your supervisor" but realized that was ridiculous.

"Solomon Chu. Chu is sometimes Zhu, but in what's left of spoken Chinese, Chu, is more accurately pronounced, szhoo." He smiled in mock pride. "It's the surname of the emperors of the Ming Dynasty. Not so important anymore; very few of my ancestors or relatives remain. I am close to the last of the line."

Carl protested. "But there must be a million Chus in China. I was on Changzhou last year and the name is almost as common as Smith or Jones in Kansas."

Chu blinked and asked Parker, "How much have you told him of your tour yesterday? I thought you were given a pretty good idea of our condition. Did the memory wipe take the history and not take the rest?"

"Uhm, Carl." Her attempt at mood lightening died. "There may not be that many Chinese left. There was a war."

"Nuclear?" Carl's face betrayed his understanding.

Parker nodded. Chu left the room and busied himself with a terminal. She cursed herself. *Dammit, I should have told Carl a bit more.*

"I'll make it short." Noel picked up the explanation, looking over his shoulder at Chu's back. "Your environmentalists in the 21st are banging the gong daily about global warming, sea level rise, and inevitable catastrophe. Your business and political leaders, especially here in the US and what you still call the developed world, are looking to maximize profits and not much more. Am I right?"

Carl nodded, mouth grim.

Noel continued. "Well, the planet continued to get warmer while food shortages, coastal flooding and failing petrochem output, and massive crop failures, all exacerbated ethnic and religious conflicts going on almost everywhere. Meanwhile, trillions of the world's GDP was wasted on colonies on the Moon and the Mars outpost."

Noel paused, took a sip from his tumbler. "The jihadists started it over a land squabble between Bangladesh and India. Israel took out the Kaaba. The Russians tried to take out Israel, but their missiles took out half the Middle East. NATO struck back, China hit Russia, Russia hit back, and when it stopped, most of the planet was radioactive."

"It went pretty much according to script," Carl said. "That's one of the scenarios a client of mine came up with."

"MAD," Parker mumbled.

"Yeah, crazy," Carl said.

"No, M-A-D, mutual assured destruction," Parker corrected him.

"Right, old school sixties war planning."

Noel said, "It took a hundred more years for everyone to forget."

"Jee-zus H. Christ!" The syllables came out in rising crescendo. Carl was visibly losing it.

"Carl, please." Parker had had more time to adjust to the reality of a lost future, or a future loss. Accepting that had been part of accepting the ultimate weirdness of a yester-today in which she was here for a second time and a participant who had been here yester-today wasn't here today.

Noel looked behind him into the next room. "Come on in and sit a while. I'll get you both something to drink. Too much bad news can mess with your gut."

Parker and Carl followed Noel into the second room and took offered seats. Noel took two stainless tumblers off a shelf and filled them from a simple spigot in a recess. He set them in the center of a little table. Carl and Parker looked suspiciously at the cups, then at each other. Noel saw the interaction and smiled knowingly. He poured from one into a second, from that cup into a third and then restored the level in the first cup. He then took a sip from the middle cup. He raised it in a mock toast, "Drink up, it's just water!" He grinned at Parker, "and I just cleaned the filters."

The two visitors took their sips, cautious at first. The ultra-pure clean water tasted good. But Carl wasn't calmed in the least. "Everything I do, whatever we do. Our generation will be wasted. It's all for nothing. It's crap, I'm crap. The whole world is going to go to crap?"

He spun on his seat to face down Noel. "I'm not crazy right? If everything you've both told me is true, you guys really are from the 23rd Century, and there's less than 600 million humans left. That's about how many of us there were in the freakin' 16th Century!"

"That's about right. And most of those 600 millions are living in a tribal or feudal post-technocracy dark age with no ability to manufacture the things that are breaking down all around them." Noel was not apologetic. "And if we weren't still on a down spiral, we wouldn't be here." He raised an arm, pointing at the walls of the time station.

Carl, only slightly mollified, "Still on a downward spiral?"

Chu spoke up from the adjoining room. "Ya manno, we din' take dis vacation for the sunshine."

Noel shook off the interruption. "Yes, we're still on an overall downward spiral. Too many of our survivors are barren or sterile. Life expectancy is down. Each generation is smaller, we are fractionated by life on disconnected islands. One thing you two need to understand, and I don't think I've made the point yet. This did not all go bad because of the Last Day."

Parker and Carl exchanged looks; their shared expectation of more bad news was not unfounded. Noel continued, "OK, I'll share the long version. The road to the Last Day was a long, hard road. You guys, and your parents and their parents, going back to the 1950s. You all really screwed up big time. You had warning after warning, and no one listened. Political hacking by special interests led to a long line of do-nothing assemblies here in the US, in France, Germany, the UK. Even worse in Russia and China. When they came on the world stage vying for lead suppliers of goods, the US slowly sank to third, it really got bad for the Third Worlders when the US pulled their military out of harm's way."

"I could enumerate the bad decisions, but there are way too many. Most of the First World and the emerging Second World powers knew about the runaway greenhouse effect; knew about the warming oceans, melting polar ice, loss of the Greenland and Antarctic ice sheets." He raised his hands in frustration. "And nobody did anything significant. By the time the southwestern US and the Arabian Peninsula were covered in solar panels, it was too late. Those things should be there now, not fifty, sixty years from now." He nodded toward the workstation where they'd seen the videos. "Those flooded lands you looked at displaced millions upon millions of refugees that nobody wanted. Not even in their own countries. War, conflict, drought, famine, disease. Buddha's balls! And it wasn't just the poor. The southern reaches of Florida displaced over five million. The coastal flatlands of Northern Europe almost as many. China's lowlands displaced thirty-five million. The jungles of South America, India, Central Africa? Gone, burned for firewood."

He shook his head. "Hindsight, it's a real kick in the balls." The kids of the 2060s really had it in for the so-called Millennials. That was probably the last generation that could have made meaningful change. Did they vote? Did they go out and change things?" He took a sip of water from his tumbler, banged it down. "Nah! People are basically narcissist xenophobes at heart. Leave you hands offa my stuff is an over-riding character trait."

Noel barked a short laugh. "Gets really nasty when there's no more stuff, I can tell you. So, don't go away thinking this all went downhill in the 22nd. By then it was all already royally bolluxed. The Last Day just put the icing on the cake."

Parker found herself wiping a tear track dry and looked over to see an ashen gray wash come over Carl. "Carl, are you OK?"

"No, feeling a little off my breakfast, I think."

Noel offered to get Carl something for his stomach. "No thanks. It's a bit much. I mean, we've been hearing the warnings and all. Seems kind of remote. A centimeter here, an inch or two there. Seemed like the beach would just adapt."

"Sure," Noel leaned back against his chair, tucked an ankle up on his knee. "Think about a meter before the end of the 21st, and then it sped up. Engineers couldn't keep up. Politicians — " He looked off into a vacant space. "Don't even get me started on the politics. All those big and little countries trying to be in charge, positioning for head of the line in a negative sum world."

Noel looked at Parker, eyes narrowed. "You have any kids?"

"Umm, no. I can't."

"Good on you, you won't have grandkids going hungry. Fighting for scraps."

She stiffened. "That's a horrible thing to say."

He shot back, "It's a horrible thing to watch."

They sensed a need to back off a bit, and each in their turn took a drink of the CalStation's purified water from their hi-tech heater/cooler tumblers.

Carl, recovering, asked, "Noel, what's left? You guys are here data mining tech formulas and technical how-to manuals, right? White papers? Patents? Are you at all sustainable?"

"I could go on, listing, but many of the technologies we're losing, you don't even have yet."

"Noel," Carl asked, "do you need any help, in the here and now? In your 22 whatever, or back here in 2020? Anywhere in between?" He shared a stare with the dark-haired apostle from the future. "I'm essentially unemployed right now, have sufficient means to not work for a while, and would be happy to help."

"Sufficient means? Tell me, what does your present financial standing have to do in light of what you've learned here? Relevance, please."

"Relevance? Well, for starters I am, without undue modesty, adept at international data flow. I am, or was, an information broker for the rich and powerful. Some or our biggest corporations have secrets they'd like to protect, and I help secure the front door. Extra locks, et cetera."

Parker looked sidelong at him, impressed. Chu, turned from his workstation, head tilted in consideration, and said, "Really?

Your skills might prove useful. Are you as good at offense as you are at defensive systems?"

"If you can build a lock, you can open it," Parker offered.

Carl said, "I've built some of the best system security locks out there. Depending on which doors you want to open, nobody would even know I was inside. Every system out there has a back door. I have the key to lots of them, no matter how hyper-encrypted. I encrypted them."

Chu nodded understanding, swiveled on his chair, tapped a screen to life, and began navigating a file system by moving several rectangles and triangles around, covering certain portions that displayed in different colors as they overlapped. Watching, Carl thought he might be doing a multi-lateral search. Chu drew one of the boxes out of a resultant set. Satisfied, he triple-tapped the forefinger to thumb, and a printer slid out a sheet of thin paper. Chu looked the list over and handed it to Carl. "Have you created your 'door locks' for any of dese?"

Carl scanned down the list, as he read down the page, his eyes widened, and cheeks puffed in preparation for a silent whistle. He opened his mouth, ready to speak and closed it. He finally said, "These are some of the planet's largest corporate entities."

"Well? Have you?" Chu challenged.

"Yes, a few of them. And they operate with complete disregard for any of the social or environmental damages they might cause."

Parker peeked over the edge of the paper. She did whistle. Addressing Chu, she asked, "What do you have in mind? For that matter, what exactly is your mission here, or there, or in the 21st?" And finally, "In my time?"

Noel took the sheet and read the list. Chu leaned into the back of his chair and rubbed his hairless chin between thumb and forefinger, then began to gnaw at the ragged edge of a fingernail. The four exchanged glances as if in silent interview. Finally, Chu asked Parker, "You said you were a lawyer woman. What kind a lawyering?"

"Family law mostly, some wrongful injury. I did corporate for ten years, before I went out on my own, but my old do-not-competes kept me out of that line, and now I've got a local reputation in the divorce and custody courts."

"How good were you at corporate? Were you into trademark, acquisitions, turf wars or what?"

"A little M&A, mostly turf wars."

Carl supplied the explanation, "M and A, mergers and acquisitions."

Chu's mouth thinned in a blend of mischief and challenge. "Bein' a lawyer lady, should I consider dat your personal sense of ethics is situational, client-based, monetary, or should I not assume?"

"I'm not going to say that I'm amoral or immoral. I guess I've let a few clients go who were personally disgusting to me. People I wouldn't or couldn't represent. They come up, you know? Serial child molester in a divorce setting. Can't do that. I caught a piece of a merger case a while back where spurious tech crimes were laid on a start-up that had a very good line on some 3D printer technologies. After the acquisition of the start-up, I found that the shark that ate the guppy planted the evidence on the guppy's hard drives. That really pissed me off. That's what made me want to leave corporate. It goes on all the time."

Chu raised his hands in the stop gesture. "What I'm tinking is, that you two canna go back quiet like an pass off your Sunday morning adventure at a dinner party. A weird tales episode dat you know no one will believe."

"Well, I'd really like to go back…" She reached for Carl's hand, squeezed it.

"Ms. Parrish, you seem particularly resistant to our memory blockers," Chu said. "Not sure dat heavier doses might not cause permanent damage to dat lovely brain. I tink I don' have any choice but to induct you into our little club. If you'll consider it."

"Cool!" Carl clapped his hands. "When do we start? What can we do? Is there training?"

"Not so fast, manno." Chu again put his hands up to stop him. "Our cybers work systems so much more advanced than you've been working on. I'm sorry, I need a lawyer in your time, not systems engineers." To Parker he said, "Remember, I was going to ask my superiors about you yesterday? Well, they're open to recruitment on a very limited basis."

Noel had stepped into the tiny room that looked to be sleeping quarters, but turned back toward them, mouth open.

"Chu, we talked about this. Wellington's permission to interface was limited to simple interroga—, interviews."

Chu waved his partner off, his fingers flying in a series of flicks that might have been addressing the motion-sensitive computer systems. Noel read the sign language and turned away.

Carl was not going to be minimized. "But there are system locks that I and others in my business put in those off-the-shelf security packages that your people won't find. Triggers. You may be able to screw things up, but you'll leave tracks. There are reasons for not leaving tracks."

"Convince me." Chu was impassive, stone faced.

8

Carl read down the list, his moving index finger stopping a third of the way down. "Here, Abadon Industries. They have major headquarters in Topeka, Kansas. You can get into some of their world-wide operations, get some data dumps, but if you want to do some serious damage to the Abadons, you need to be inside their server rooms in either Topeka or San Francisco. From there you can get into the banks in the Caymans and Zurich, get a choke hold on their purse strings. With those purse strings, you can finance, in local 2020 dollars, a whole lot of mischief."

Chu had been chewing his fingernails to a nub. He didn't like new problems, and these two visitors were new problems. He stared back at Carl hard, gears meshing. Maybe he'd lucked into something with this intruder. The young tech wizard had the benefit of credentials. They wouldn't have to create the essential documents and then hope no one who shared the resume in local time and space would cross paths. Could Carl actually be an asset? Chu wanted to hear more. "Go on."

Carl paused to think about a few details; he was rapidly crafting a battle plan. He'd done it before, working in the big leagues. "The Abadons have annual revenues of over a hundred billion dollars. That's annual and that's billion with a capital B. That might not be a lot of cash in 200 years, but here in 2020? It's more money than most of the Third World's GNP combined. Sidelining a few hundred million into a spare public interest group that focuses on research and development will be no worse than an accounting error. Some narrow-tied suit on Wall Street will get fired. No biggie. Parker here," he waved a hand toward her, "could set up a registered California not-for-profit political action committee. Our government is very, very bad at following up on the activities of PACs." He let his grin widen, pretty sure he was on

73

a right track. "You could do a lot in 2020s with a few hundred million."

Chu did not smile. He simply stared. His usually neutral face was a study in concentration. He took them both in, their apparent eagerness to help. Would they be prepared to take the effort to the next level? Could they live with a mission that was designed to rewrite history before it became history? His own supervisors in New Zealand were not on board yet with his reassessment of the revised timeline.

Chu's grandparents had fled their homeland and made it to New Zealand. Prevailing winds from the east meant minimal collateral damage from any destruction on the neighboring Australian coastline. The last leg of their exodus took them from the hard crash of Hong Kong's economy to New Zealand. The company's helo had just enough range to make the hops from Hong Kong, to Manila, to New Guinea, and Brisbane and finally to a pre-purchased refuge on the North Island.

By the time they'd arrived, so many refugees of all origins had made it to the Islands and so much of the remainder of the world was radioactive or gone that the locals had come close to reinstating the infamous New Zealander prohibition on immigration of non-whites.

Being Chinese had not been easy for the young Solomon Chu. The rural setting of the North Island and its new population of refugees had made the multi-ethnic population of the swelling village of New Tokyo interdependent. But after two generations, racism was still present in the original whites, in spite of the technological prowess the immigrants had brought to the emerging world capital of the 23rd century.

Old enmities fade slowly, and the majority immigrant population of Japanese had made sure that life wasn't easy for a small Chinese boy. He'd taken refuge in a ganja street gang where he befriended Grieg. Soon, his street cunning, obvious intelligence, and calculated political skills elevated him to the top of his recruit class in the service. His body training and aggressive skills in wǔshù boxing helped in the rougher spots.

His wizened eyes appraised Carlos Reyes unsympathetically but hopefully, Carl was right. A few hundred million dollars could do a lot. But if he could cyber snatch a fortune, he could do a lot more to disenfranchise those at the top of those fortunes. Carl had begun to fidget under the stare. Chu relented, "All right, Carl and Parker, welcome."

~ ~ ~

The drive back over the mountain and down to Orange County was charged with an odd mixture of sexual tension, depression, and excitement. They had stopped near the peak and watched the sun slide into the Pacific. Barely touching, holding hands, they watched in silence. Their conversation, thin on substance, was full of new meaning. It was one thing to believe that all are mortal, and death is the natural outcome to life. It was entirely another thing to think that it will all come to nothing. Not in the next fifty years, but shortly afterwards. Any family they might have adopted or made; down the line, any children and grandchildren would either be incinerated or poisoned or, worse; survive humanity's penultimate paroxysm to experience the chaos that followed.

Chu had described desperate scenes of scavenger hordes preying on the small freeholds of still-civil survivors. Shortwave radio calls for help from outposts in dozens of languages had been recorded for posterity, but Parker had only been able to listen to two of them. Chu had insisted; she resisted. There were no movies of the actual war. A few distant shots of mushroom clouds had survived, of course; there was the vid from the weather station but little else to convey the near totality of the holocaust. Chu was complete and unsympathetic in his descriptions of the fall. When they stepped back out into the warm Sunday evening in March of 2020, shock set in. Carl drove slowly, almost with solemn attention, as they took the dirt drive back over the mountain.

They'd gotten to the top, to the little pullover lookout made by hundreds of others who'd stopped to look at sunrises over the valley or sunsets over the Pacific. A dreary certainty settled in on them both. Parker started sobbing and reached over to find him withdrawn. With awkward jerks, he pulled his shirt sleeves to dab at his cheeks. Silently they both got out, walked to the front of the car, and embraced. Her sobs had turned to body-shaking grief. His tears melded with hers. All other emotions were subdued by sorrow. How do you grasp the reality of the end of the world?

"Carl, how can I get my head around the idea that humans, doing what humans do, will eventually allow short term financial greed, religious dogma, and tribal jealousy to bring an end to civilization?"

He gave a barely audible grunt in response. "We are destined to give the whole world back to the cockroaches and rats. If this reboot that Chu and his team are attempting fails, would humanity eventually win again? Or the cockroaches?"

"You ever see Mad Max? Or A Boy and His Dog?"

"Oh God!" She shivered, took his hand again. "I'm thinking more of one of those sci-fi fantasies where a toppled Statue of Liberty is drowned under a hundred feet of seawater. Um, *AI*, something like that."

As the swollen red ball of the westering sun kissed the horizon, he lifted her chin and kissed her. They talked the sun down, considered the uselessness of planning. How do you plan for extinction? In time, in the growing chill of a moonlit panorama, the conversation turned to how do you *plan* to avoid extinction. It couldn't be enough to simply live on without taking Chu's offer seriously.

"Carl, we *can* do something about this."

"I've been thinking the same thing. I'm pretty sure I can get you the funds to start up an environmental action group."

"I sure as hell know how to set up a not-for-profit corporation." She took her eyes off the darkened horizon, shivered, "You want to be my partner? Executive Vice President in charge of cybersecurity and chicanery?" She asked, her tone expressing a new, lighter, maybe hopeful mood.

"Oh, so it's to be Vice President? I can think of some vices right now." The mood had changed. He pulled her to him and kissed her.

She returned the kiss hard and full on, and when she reached for his shirt collar, he took her hand, lifted her to her feet, and guided her to the car. "Let's go home, babe."

They drove down the southern slope to Pyramid Lake in silence, talked out, emotionally damaged. In the drive through LA County, she outlined her immediate plans to close her practice. She'd have to give cases away to colleagues, cancel any outstanding appointments, and find a place for Meg in their new start up, whatever it would be. Meg had always been supportive;

she owed her the option if she'd take it. They decided he'd move into her house. He'd maintain his home office, his tech center, but moving in now seemed right. Besides, her infinity pool had excellent views of Pacific sunsets.

~ ~ ~

Parker awoke on Monday morning at 6:41. Dawn's earliest glimmer profiled the mimosa tree outside her bedroom window. She'd been dreaming, but not of a 1930s farmhouse. As the dream scenes of a serpentine blacktop faded, she felt the slight pressure of a grin on her cheeks. The farmhouse dream was gone.

"Way to go girl; you slept the sun up." Carl's voice filtered into her consciousness. Carl! She rolled over to face the usually empty other side of the king-sized mattress. Nothing!

"Over here."

She turned again and saw him curled cross-legged in her papasan chair. His fingers were poised over the keyboard of his laptop. "Juice? Coffee? Tea? Me?"

"That offer for coffee real?"

He picked up a large mug and waved it toward her.

"Then yes." She propped herself up on her elbow. "Have you saved the world yet?"

"Not yet. I've set up some history, though."

"History? I thought we were going to fight the future, or for the future."

"I've been setting up some routing so that anything I do on your Wi-Fi node looks like it's being routed through Manchuria from a variety of servers in Silicon Valley."

"The China Manchuria?"

He let out a small laugh, "There's only one that I know of, and they have one of the best hacker schools in the world." He took a small sip, slurping the hot liquid. "I know a few of them by their signatures. So-Called-Wizards has some history solving Third World problems," he said, as a mischievous sparkle danced across his eyes, "and causing a few First World problems.

He stood up. "We're problem solvers and you need coffee. Why don't you do whatever you do first thing and come into the kitchen for breakfast." He sat on the edge of the bed and ran his eyes over the lines of the sheets as they draped her hips and thighs.

Parker had been resting on her right side, propped up on her elbow. She gave her left shoulder a subtle shrug and a loose fold of the bedsheet fell from her shoulder. She felt a stiffening in her nipples as a stray eddy from the AC chilled her newly exposed skin. She smirked as his eyes struggled between eye contact and appreciation of her breasts. She could tell he was conflicted: breakfast or morning sex. Smiling, she cupped a breast in her right hand and pulled lightly on the hardened nipple. He gave up, put the laptop down, and crawled onto the bed. They unanimously decided to delay breakfast.

~ ~ ~

Jazz jumped into the space between the two humans and meowed his displeasure. He took a tentative sniff and lick at the sheen on Parker's arm, jumped over her, and leapt onto his alternate perch on the nightstand.

Carl, still cooling down from their amorous morning workout, sighed. "I guess I'm going to have to get used to sleeping with a cat."

"He's not used to you yet. When you're accepted, plan on waking up with him curled between your legs."

"I'm not sure I want him to go there." He took another deep breath and launched what had been on his mind before she kissed him fully awake. "Park? Yesterday, Chu and I laid out a financing plan that seemed almost like a joke at the time, but Christ! At best, the consequences of getting caught could mean twenty to thirty years at Otisville."

"Otisville?" She rolled to her side and propped up on an elbow.

He glanced over, saw one breast peeking out of a sheet, reached over and covered it. "I'm too easily distracted, my dear."

"Tired of me already?"

"No, it's just that we're about to do something radically different, and — " He took a breath. "Otisville is a white-collar prison, for white-collar corporate criminals. But that's a best case. I'm not famous. At worst, there'd be twenty or thirty years in

Soledad, which might as well be a death sentence. That's some scary shit."

"Wow, yeah."

He waited for more of a reaction but heard only a deep breath and the exhale through pursed lips. He expanded on worst case. She had to know. "It would extend to you, as a co-conspirator. We'd both go down if caught."

"Can we not get caught? Is this above your pay grade?"

He considered that. It wasn't a challenge; it was a valid question. "With the right targets, I know I can get inside. Keys to the city, back door, whatever you want to call it. There are some big accounts I can get into with no trouble. Of course, there's the dark net too."

"Dark net? Should I ask? Sounds like conspiracy theory BS."

"It's there, an elite group; you never know who's who. Some are actually Feds, some are cops. Most are just guys like me, professionals in the security world who occasionally step out of bounds."

"You ever step out of bounds?"

"Uhm…" He ran the list and decided it wasn't a very long list. "Yes." He had been tracing circles in the sheet with his index finger. He stopped and looked up. She was quiet, eyes closed. "You in there?"

"Yeah, I'm here." Her eyes popped open, lips in a half smile. "Carl, I know we have a very short history. So short, I'm not sure you can call it history." Her face scrunched, then relaxed. "But if you're in, I'm in. Chu trusts us, I think. We wouldn't be back here with total recall if he didn't. With no chance of bearing children, I'm pretty sure he'd consider me slag. Since we've proved the memory wipe cocktail doesn't work on me, he could have just disposed of me. Or drugged us both and done who knows what."

He reached out for her hand. Squeezed it. "He could have dropped us out the back door, into whatever we'd find in the 23rd."

Her eyes widened at the prospect. "Yeah."

He lay back, staring at the blankness of the ceiling. "Not sure I want to think about it."

"So, are we in? In with Chu?"

His pause stretched into uncomfortable silence before he answered. "Yes." He took her hand, opened the fingers, and nibbled at the web of flesh at the base of her thumb.

"Horny bastard."

~ ~ ~

The first steps were relatively simple. Parker would go to the courthouse and file incorporation papers for a not-for-profit LLC. Carl would start to lay tracks for a mythical cyber group, inventing hacker personas and fake blogger dialogs using both of their laptops, and then he'd have to dispose of and replace the laptops. Each cyber-attack would be routed through several blind addresses. One hacker bragged to the other about his back-door attacks on some mega-corps, the other daring him to prove it. The first was to move a million bucks from one of the megacorps' Cayman accounts to his own Swiss account and then they'd split it. That conversation existed solely to be leaked in the near future to throw off the security teams that would surely go looking for a lost million dollars.

"Parker, I need a name. My brain's gone dead. Cyber geek name."

"What did you say your two hackers' names were?"

"ParagoN99 with a capital N and Gr8efX." He shrugged.

As good as any, she thought. Then, "Try EyeCandy. One word, but with capitals. Can your cyber thief be a hot chick?"

"Yeah, that's great." He typed out "ICanD" then grunted displeasure. "That's a little too short for an ID."

"What's the gee-eight thing about?"

"As I've sketched this one out, he's supposed to have risen to fame on the boards for having crashed the servers at a G8 summit and leaked strategy emails from Chancellor Merkel and sent them to the Greek and Italian side when they were debating EU politics."

"Really?"

He laughed, "No, hell no. Every bit of this is bullshit. All leads are designed to go nowhere at all but to serve as serious distractions. So, FrICanD, the hot chick hacker is French."

"Damn, ever thought of writing books?"

"No, that's a long-term commitment."

She looked at him, lost in his imagination, creating bad guys with history, creating lost entangled threads in alternate Internet histories. He was bent over his laptop, typing at speeds that made

Parker a little jealous. She hadn't really gotten proficient at typing until grad school. At only half a generation younger, Carl had grown up with a keyboard in his hands and could text, check emails, track stock market prices, and be a cyber-terrorist all at once. She felt happy to get a brief typed up so Meg could wordsmith it in time for filing. *The courthouse!* "Shit!"

"What?"

"I have to go in. I need to go to the courthouse anyway to file these papers, you still have to go in and sign a real piece of paper. And I need to go to the office and begin to clear it before the rent comes due. I have to say a few formal words to people who've been good to me there. Make sure that my client list will end up in good hands, and kind of shut down that chapter."

"You're keeping Meg?"

"Actually, yes. I hope so if she'll come. She's a green-tea-drinking, solar-powered ex-hippie. She'll love this."

"What about your other associates? Are they likely to get suspicious of you leaving suddenly?"

"I don't think so. I had actually been talking with James — he's a medical malpractice pro on my floor — about taking a break, taking the trains through Europe, wine tasting, maybe taking a room in Florence, and getting into the Renaissance art scene. And besides, suspicious of what? Starting a new pro-environmental corporation." She waved the California Corporation filing papers in their manila envelope. "And maybe we really can take a wine-tasting tour of Europe."

"Excellent. That actually would be great cover for some 'missions.'" He closed the quotes with his fingers.

She began to pack up her very small clutter and folded the top of her computer "Meet you here for dinner? We can toast the creation of the 'Room for Tomorrow' environmental good works not-for-profit corporation."

"Room for Tomorrow?"

"Yes. A little room in the upstairs of a forgotten country house. I think I took you there yesterday. Or was that yester-today?"

"Hmm, Room for Tomorrow. I like it."

9

Noel Antonides stepped into the decon booth, feeling dirty and shut the outer air lock seal. His body relaxed, finally. Being outside in the hot, hostile afternoon sandstorm and the attendant horrors of unseen predators was one of the worst aspects of their assignment in CalBase. The time station was a convenient artifact, but it was an artifact. It was decades old, abandoned by a generation that had complex but simpler problems than his. *How much longer are these scrubbers going to actually scrub with a limited supply of filters?*

The usual sense of nausea told him the other door was set to the 21st. Were the visitors back? He waited while the air cycled. The hard blast of air jets cycled slowly down his frame, blasting first helmet and shoulders and working down toward his ankles. Any micron or larger particles blasted loose by the jets were whisked into gel capture. It would not do to contaminate the 21st with 23rd century air. As bad as the smog was on the worst days in the 21st, it beat the hell out of a slow death breathing the contaminated Californian atmosphere of the 23rd. Exiting the detox chamber, Noel called out, "Honey, I'm home!"

Chu called from his workstation. "*I Love Lucy*, 1950s."

"Damn, thought I'd have you on that one."

"How's d'air."

"Still breathing." Noel coughed once, joking.

"Go on in and link to de filter banks. Chat ya when ya come off."

Noel noted the data cube on the input pad then looked at the screen glowing at Chu's workstation. "Anything interesting?"

"Cha. G'now."

Noel pulled a bunk from a recess in the wall and sat down on an edge that allowed access to the filter bank. He rolled up his sleeve, exposing the IV implant. He flushed the port with an antiseptic, carefully unwrapped a link, and plugged himself into the blood filters. He tapped the control panel for a sleeper dose and lay down on the bunk. When the shot of sleeper hit, he relaxed and was under in moments.

Chu closed his screen down and came in to check on his lead investigator. He sat on the edge of the bunk and absently, lovingly, straightened Noel's shirt at the neckline. He checked the link connections to ensure all the connections were sealed and then checked the numbers scrolling on the monitor. Satisfied that everything was in order, he stepped into the shower to freshen up.

He was out and dressed again when Antonides's blood filtration was complete. An electric tone indicated that a wake charge was entering the drip stream. Noel slowly came to, shook his head to clear it, and sat up. Chu was smiling from a chair opposite. "How did you do?"

"Wotsay manno? Doon 21st?"

"Cha, Noel, I need de practice."

"O'fine." Noel recalibrated and replied in 21st. "OK, fine."

"So, how'd it go? Did you get in?"

"Yeah, the encryption wasn't anything we haven't seen before. I set the descrambler on it and it got in before the room got down to temperature."

"Servers intac?"

"Some initial boot problems. I figure they haven't been turned on since this area was flashed and they lost power."

"Chick-a-Saw built great cloud banks in de day. But so much for small talk. You pull any data down?"

"Not only did I pull down what we were looking for, there's a gold mine in there."

"Yeah?"

"Bot damnation!" Noel's grin split his face. "We'll have to wade through piles and piles of shyte. People never throw any pictures away. There're pictures of thumbs, the ground, nothing at all, and they kept that crap in the cloud for me to sift through — and the porn! Buddha Balls! Zettabytes of the stuff."

"Sure, bet you hated goin' tru dat!" Chu screwed his lips sideways. "Get any ideas?"

Noel blinked back in mock disgust.

Rebuffed, Chu turned on a barefoot heel and moved the three steps required to enter their compact kitchen.

Noel ignored him and his leer, and turned to his workstation, dumped out data cubes and began sorting them. His mission had been simple: after five weeks of searching, he and Chu had located the remaining server farm for the Chick-a-Saw cloud storage facility. Now that they had gained access and booted the rack, he had a list of clients to search for who, before the Last Day, had their corporate records routinely backed up onto Chick-a-Saw's rack storage. His biggest obstacles, aside from the toxic environment, were the roaming bands of renegades who occasionally came into the radioactive lands to forage for parts and the characteristic dangers of the mechanical death of the region. Second to those were canines that had the pack hunting characteristics of coyotes with the bulk and meanness of German Shepherds. On good days the packs shifted from shadow to shadow, on bad days, they could suddenly appear out of the clouds of blowing dust. Having zappers helped, but sometimes the sheer numbers of them were harrowing.

In the underground server banks, almost everything had been protected by their hermetically sealed clean room environments and their EMF resistant Faraday cages. Anywhere else, equipment was rusted beyond use—motors frozen, hinges locked for all time, rubber rotted to powder. Vehicles had not traveled the winding interstates in decades. Despite the washouts on lesser county-maintained roads, the interstate to the east was technically usable. But anything on it in daylight was little more than a moving target for bandits.

The eastern fringes of Neenach City were frequented by foraging teams. Although lateral blast waves from over the mountain rendered must structures unusable, the valley had not taken the direct hit damage experienced by the more densely populated western slopes of the coastal range. Some items in original packaging were still usable and worth an expedition into the toxic environment. Foragers from across the Antelope Valley came into the ruins of Neenach well-armed and deadly. Like them, the few surviving bands on the continents had found that civilization was a hard thing to maintain when global trade was no longer available.

Chu had been watching Noel work and knew his partner was conflicted about the possibility of help from the tech wizard and the lady lawyer. He decided not to bring it up; the afternoon had possibilities. "So, you found the client list?"

"Yes, but I think Grieg is going to have problems." Noel did not sound optimistic.

"Why? Where did you see him last? You left him down there? You gone back?" Chu was worried now. "Wot about Grieg?"

"Not to worry, mate. He took the other cart."

Chu felt uncertain, strange. It seemed like Noel had gone through some weather change, or he had. "So manno, where is Grieg, xactly?"

"He was in the power supply room when I left, trying to see if he could restore the feed from the building's PV array down to the server room so we wouldn't have to lug juice boxes with us. Someone had scavenged the conduit from that run, and he was trying to replace it with jumpers from the internal circuits."

"Too bad. Dose power feeds were carbon composite." The realization that this was exactly the kind of research tech the 21st could use, he reconsidered talking about his plans. "We need to talk about Parker and Carl."

"Cha, not a good plan, manno." Noel turned slightly to return Chu's glance, but returned to his work.

"I tink, dey help on their end." Chu wondered if this was the issue. Was it the offer to take help from the visitors?

Noel pushed back from his workstation. "I don't want to argue over it forever, manno. You're talking major mission creep. We've only been in CalBase a little over a month, rebooted its systems, found our data targets in the valley." Chu stepped back as Noel rose from his chair. His friend's face was on the edge of anger. "We are just now getting to the serious work of our mission, and you, you're pulling in slag from the past."

"I tink we can use Carl, he's got high-grade 21st net skills."

"Wow, high-grade 21st, huh. That means he'd be able to do what exactly, write a mobile app? Replace an optic drive?"

"No serious, manno. He seems to be a hacker's hacker. Known in the trade. I tink I can use dem to pursue my side project."

"But you never got that approved." Antonides was surprised to hear that Chu wanted to revive his "side" project. It had been discussed in the governing council in Wellington and they were waffling on approval.

"Dos sorry bastards are 'fraid of temporal parallax effects."

"Well, that's a given, isn't it?"

Chu began chewing the nail of his index finger — one of his tells of nervous energy. He said, "I don't see any reason why we shouldn't change life as we know it." He spit a sliver of nail into a waste can. "Yesterday was, for them, Sunday, the same Sunday she visited the day before. She came back at essentially the same time, with a friend this time, and didn't find herself, or tempt the Gods' timekeepers."

Chu could see from Noel's narrowed glare that his friend's irritation, his stress, was real. He considered the recent past, the mission's initial difficulties. Their original expedition leader had died of toxic shock poisoning. He'd torn his suit in the perilous crossing through the blasted ruins of what remained of Los Angeles County. Now, he and Noel were more or less equals with different areas of responsibility and some cross-training. He knew that Noel did not like a democracy; he liked the regimen of the service. He would be more cautious of the implications of a change in local time events.

As if reading his thoughts, Noel broke the silence. "Chu, if you're successful you might cease to exist. I might cease to exist; we might never have been born."

Chu was truly pained. "Noel, our world is a bleedin specter! Damore I look at de 21st, the more I realize dey had some 'mazing achievements. True, dey were already on de downside and didn't know it, but dey had a culture rich environment. De global satellite system was really remarkable. By comparison, we barely have a shell up. They had almost 100 percent comm-sat coverage, any bleedin' where! We, we have chicken squat in comparison. We still have to aim a bloody dish at the West-Pac to link with Wellington."

Noel shot back. "Yeah, well they still died of 400 kinds of lung diseases from simply breathing."

Chu sat down across from Noel, facing off. "Sabe, and dey din really enjoy being old no matter how long dey lived cause dey couldn' delay cell necrosis. But dey had a wide world to play in, and der children were vital. Ours, hell we're lucky when a dozen of ours can germinate one ovum."

Noel took a moment to respond. "Chu, they had their shot at history – they blew it. It's our turn, and they left us a real bleedin' mess! It's like you're rooting for the Romans. Crikey, they were great too, if only they didn't murder for fun and games on Sunday afternoon." Noel stared at his partner. "Listen, mate. Why not grab a cuppa and think about it? You've gone thick!"

"Dose Romans also prayed to virgins and da wedda." Chu stood and began to pace. "Hear me, Noel. Dere's anodder way. We can course correct de past and save de incineration of two billion people an de poisoning of almost everyone else. Hell mon, my grands died in their fifties of leukemia which dey caught from breathing! Breathin! Air is supposed to give you life. It is not supposed to kill you! Sabe?"

Noel stared hard at Chu, now nibbling his left thumbnail. Their relationship had become complex years ago. Competitors, then friends, then partners. They had contrived to share this assignment, but now this. Power struggle? Disobeying orders from Wellington? Bringing in slags from the 21st. He said, "It's not like anything they do will bring back the 22nd. Hell, the 22nd progressed a lot further in techs or we couldn't be here." He turned around again to face Antonides.

"Yeah? Look at this room if you want to think about losses." Chu gestured at the enclosure. "Dis room is a bloomin' time machine!"

"Cha, and you go bollux it, and we all go pfft." Noel crossed his arms, closing down, and dropped into street patois. "Oh, so sorroh, you wan ge-yo practick spekin' in da 21st. You goin walkabout in the 21st like me, like Grieg. Mebee yo don wan go home to da 23rd and all its dismal opismal problems, but manno, it is a brave new whorl we gone buill. My gran, your gran, tay had the get-em-up call. Now we, we starting over!" He stood, nose to nose now, arms crossed, not sure whether to land a punch or give a hug. His chest heaved as he cleared his head and some of the anger. "Chu, we just need to get some more how-to-do-it secrets from the tech boxes and go back to NZ. Mebbe we can adopt a kid, be a family!"

Chu seemed to relax; arms now limp at his sides. "Noel, my dear man, I'm no say'n I wan'be stayin' here in da 21st. Crikey, I can barely breath dat air when da winds come over da mountain. But t'other night? When we took Parrish back. Manno, do you remember how clean that ocean smelled?"

Noel fell silent and looked away. He started to move into the bunk room, paused and turned. "These buggers we're data mining. They had everything tech could bring, wasn't perfect, but it was good, Cha?"

Chu only blinked.

"And they fucked it bad." Noel squinted; head tilted. "And these two new buggers, Carl and Parker, they and the rest of the 21st, they goin' to pull it all down on their heads!"

As if taking sides with his new friends from the 21st, Chu tried his best to enunciate in the hard sounds of their speech. He tried it out slowly. "The 21st did not—they brought it down on their great grandkids' heads. Now, most of the great cities of the world are hot spots that won't be walkable for a thousand years or livable for twice that. Any of your data cubes going to do something to put an end to tribalism? To end murder in the name of religion? To unmitigated greed? Those are the things that brought down the 22nd, and yeah, the problems were well rooted in the 21st."

Noel paused before answering, his eyes drifting along the edges of the white on stainless steel lines of the CalBase time portal. "And what have we got? Less than a percent of their world's population? Our bio-teams don't do anything except chronicle the species that are gone or going. Our physicists are just trying to figure out metallurgy again, if only they could get at rare earth metals. The E-techs have just about maintained our computing abilities, but we still have a long way to go on doping. We're limping along on equipment left over from survivor technology. Why else would we be here?" He waved his hands around the small time capsule. "And we don't even know how to make dis ting slide down de timeline."

"Timeline yo buns, Noel! Two billion people were roasted alive; five billion more gone in a year! Crikey! And those were the lucky ones? What if we could save all that suffering? Engineer a course correction?" Chu's neck veins were beginning to bulge.

"Mate! Look, you need to give up on this wild-assed plan to save the future from itself. It is what is. IT happened. History happened. Shyte happened. You go shaggin down the timeline, there's no telling, no way to tell what might disappear from our time." He stood and faced Chu, eye to eye again. He snapped a finger in his partner's face. "Pfft! Like that, you or I and everyone we know could, probably would, disappear. None of our parents would have met each other. We, you and I mate, will probably cease to exist if you make major structural changes in the timeline."

"Mate, hell. Can' you see wot we've all lost? We are a vestige, a whisper of humanity. A mere shadow of what we might 'ave been. Our potential for exploration beyond our atmosphere has been set back centuries. Humans will never get back to Mars to restart the colony." He shook his head slowly, sighing deeply. "Damn Noel, I wish you could see what a possibility we have here to make some changes. What if we could tweak the timeline? Make the suffering less, make the losses less, stay on an upward course?"

"Chu! Can you hear yourself?" Noel was backing away, frowning.

Chu didn't stop. "What if we could avoid nuclear 'nihilation? Back in NZ, we no longer have critical mass for cultural dynamism and growth. Dere's only two universities in de world now can do new research! An dey all rootin aroun' in de past like us!"

"Chu!" Noel slapped him to stop the heresy. Chu slapped him back. They both stood silent, shocked. Disappointed in each other, sensing overwhelming loss. They had both been looking forward to a private moment out in a cool evening in the 21st. Now that option lay shattered, painted in matching handprints on each other's cheeks. That possibility was over. Each knew it.

Noel broke the growing silence, "True dis Chu, we two gone longtime on."

Chu didn't have a response; the resignation in Noel's voice was heartbreaking. He turned and went into the second room. He muttered almost inaudibly from around the corner, "Noel, it's too late." Then added, louder, "Don't forget to take the rad blockers. You'll lose your kidneys if you don't."

Noel answered, so low it was little more than a thought. "Love you, Chu."

~ ~ ~

Parker's black Jaguar pulled into her driveway immediately behind Carl's red Porsche. The throaty murmur of both motors idling resonated in the walled enclave between house and her stuccoed property-line wall. They turned off the ignitions simultaneously. The resulting quiet carried an echo of the powerful engines.

Carl walked over to where Parker was getting her things from the backseat. "Parker? Is there anything in the garage?"

"Not much, why?" She fished her opener relay from her purse and clicked. One of the doors opened, revealing a nearly empty four-car garage, barren except for her Jeep and a row of shelving with neatly labeled boxes.

He pointed to the last bay that didn't have a car parked in front of it. "I'm thinking we're going to need another vehicle. Something a little less noticeable, plain vanilla."

"Really, what do you want? A Prius? Something befitting our save the world mission?"

"No," Carl walked into the open bay, his voice gaining a little reverb in the hollow space. "I'm thinking a panel van. Maybe one of those new Ford Vans with the headroom. Or the GM/Peugeot delivery vans."

"You're serious!" She followed him into the garage and reached for his head, pulled it down, and kissed him. "Why do we want a van?"

"We'd get it wrapped with some logo we're going to have developed for Room for Tomorrow.' While we're at it, we'll get municipal busses all over the major markets here, Europe and in commercial hotspots in Asia." He'd been wandering in small circles but turned back to her. "In local languages of course." Spinning on his toes, he did a full 360 pirouette and stopped, arms spread wide. "My dear Miss Parrish, coincidental and accidental tourist in the future, recent significant figure in my life, inspiration of several recent magical moments," she was smiling at his magnificence now, "and sweet seducer. Our new philanthropic start-up will soon be worth a few million dollars."

She stepped forward and let Carl's strong arms close around her. "Only a few?"

"Come." He pulled her into the house and toward the pool. He was unbuttoning clothing as he went. "There's time to catch the sunset." In a few minutes they were in the pool, with matched highball glasses resting on the western edge of the infinity pool.

90

He pointed at the low sun on the horizon. "That, my lady, is timeless. At about 4.5 billion years, it's just getting started. Our blue ball of a planet is entirely dependent on its good behavior, and it's been pretty good to us. Not too hot, not too cold. It won't swell up and eat its young for eons beyond imagining." He turned to face her. Her body, dark in the water, was backlit by the pool light at the far end. She is one lovely lady, he thought. He said, "Our invention, Room for Tomorrow, has received start-up funding from two wealthy benefactors who have chosen," he hooked air quotes, "to remain anonymous and want us to do good things. The crowd sourcing campaign that hit the internet this afternoon will break all records in its success."

"Crowd sourcing? I thought we were going to steal money from rich greedy bastards."

"Oh, we did, and we will. After we have legitimized our fiscal resources with a very public entrance into the green movement, it will be much easier to slowly transfer donations of two to seven thousand dollars from random email accounts all over the world. Of course, most of it will come from the five million dollars I stole from the Abadon brothers corporate cover account in Zurich this morning."

"Five million!"

"Million with an M!" He couldn't help the childish grin. The self-satisfied assurance behind it knew his tracks were covered by an untraceable routing that seemed to end at the desk of an Abadon Industries accountant named Tindall. Tracks that ended a block east of Main Street in Topeka, but actually originated in a laptop in Orange County, California.

"We're rich!"

"Well, we're doing all right, but Room for Tomorrow — RFT for short — a California based not-for-profit, tax-exempt corporation is going to be very well funded very soon." He raised his eyebrows in speculation. "Soon we may have to become an LLC."

"I'm the lady who can make that happen."

He slid over along the side of the pool. "You are the lady who is going to help make a LOT of great things happen."

She leaned in and kissed him on the lips, pushed away, and splashed water in his face. "But you'll have to wait a little bit for anything up close and personal. I'm hungry."

He watched her step out of the pool, and pad over to a pile of towels. He thought, A few days ago, I was wondering what I'd be doing in a month. Trying to decide whether to go to the Bahamas or Hawaii. Now, who knows? I am crazy as hell for that lady. I could be in jail in a month or so deep over my head in who knows what.

~ ~ ~

Grieg stepped out of the transfer booth, feeling good about his work in Chick-a-Saw's power room. He'd almost forgotten the blackout glasses before the decon flash popped. The usual xenon after-smell irritated his nose, but it was one of those evils you accepted with the job. His brain counted the seconds it took for the air blast to knock off the loose bits. When the auto-latch on the door released, he stepped halfway into the room. Grabbing the edge of the door he peered around with a mischievous grin. "Here's Johnny!" He held the pose.

Chu turned to him from the protein compiler and frowned.

"All work and no play makes Jack a dull boy?" Grieg offered.

"Oh right! Uhm, Nicholson. Don't tell me." He looked inward, seeing the old 2-D vid. A small child on a tricycle mumbling "redrum, redrum." Ah, he remembered, "*The Shining*, 1980s?"

"Good job."

"Thanks. I heard you were having problems at Chick-a-Saw." The protein shake was nearly finished, and Chu rattled the cup, impatient with the last slow drips.

Grieg, suddenly hungry, asked, "Yo manno, you start one for me, strawnana?"

"Sure." Chu pulled down a second tumbler and set it under the compiler's feed line and punched in commands for a strawberry-banana protein shake.

Grieg sat down at the tiny dinette. Little of the time station's furniture was movable. The dinette, which took up so much room was a bolt down. Overall, the station had more built-in features than a well-designed camper and more space saving features than a good yacht. It had to. It had been designed to share space with a primitive structure. One that had been missed by the various tremblers and quakes that shook the mountain and was isolated enough that it had been overlooked by scavengers in the 22nd. In the 21st, it was just a lonely old junker of a house—too far away from most activities to attract notice and creepy enough to discourage most of those who'd gotten curious. In the 23rd there was only residue.

They had permanently hurt only one persistent ranger. The house was on federal property and had been considered an historic site but had never been given much attention due to shrinking budgets. The ranger had reacted badly to the memory blocker's second application. Chu thought the dosage was way too high. His short-term memory function was permanently damaged, and the ranger had been retired on disability. Grieg had checked, and the guy never had children and couldn't affect generations up the timeline. He was slag.

Chu joined Grieg at the table, setting Grieg's tumbler down. He'd had a green tea and cinnamon, Grieg, a strawberry-banana. "Any luck? Noel said you'd had some problems."

"Cha mon." Grieg slurped at his straw.

"21st, please. I need da practice. We will both be needing more of it soon. I don't want to get pulled over by 21st cops and have you babbling incoherently in 23rd."

"Cha, er, Yes sir!"

"Chillma!" Chu grinned back at him. "Be cool, 21st. OK?"

"Yeah, like no problemo, dude." Grieg took a sip of the protein shake. "Yeah, actually I did. Some scavengers had pulled the main power busses for the copper cores I suppose. The PV units on the roof were shot. But I patched into the leads from the CF unit. I checked those out, and after all these years, they can still pump out electrons. Fucking cold fusion—we need to relearn that shit too. Anyway, they're churning 'bout eighty percent of the power they had new in 2150 something."

"That's good."

"That's damned good. I strung together wire from some of the room and hall lighting circuits and ran new feeds. I need to go down in the farmhouse tomorrow and see if there's still a grounding rod to use for a bus. The laminar carbon batteries are charging as we speak. I can probably bring the juice boxes back tomorrow. I just want to make sure the laminars will hold a charge before I pull the juice boxes back. Heavy sonna bitches."

"Dat's sons a bitches or sons of bitches."

"What's that mean? In 21st?"

"Hell if I know. Prickman's whore I tink."

Grieg slurped the last from the bottom of his tumbler and pushed it away. He looked around and then into the next room. "Hey, where's Noel?"

"He went back out."

"So soon? The crazy cabbage! He'll overload his kidney's doing double shifts out in that crap. I tell you wot. I don't want to go out and pull in his inert bones."

Chu looked up, staring narrowly at Grieg. "You shouldn't have to. He's capable. 'Bout you, shouldn't you hook up to de filters? Get da crap out of your veins."

"Yeah, but I was really hungry, man." Grieg picked up both tumblers and tossed them into the sterilizer. Two brief electric zits, and they were cleaned, ready for reuse. "So, where'd he go?"

"He gone walkabout in the 21st." He touched at his cheek where the sting of Noel's slap still lingered. "Noel got some tings to tink about."

Grieg looked at Chu, looking for answers that could not come from an unreadable face. He sat down beside the plasma filtration unit's couch and cleaned his arm, prepped his IV port, and linked to the unit. After syncing his wrist unit with the master control, he set it to give himself an additional thirty-minute nap and laid back. Soon his breathy snoring filled the room.

10

Sunset had come early. Gray dusk lay dark on the northern slope of the low mountains separating the bright Los Angeles sprawl from the few porch and street lights illuminating Neenach. Parker had taken to parking behind the farmhouse, even though they'd learned the park personnel rarely patrolled the ragged gravel or two rut passage that called itself road access. She and Carl got out and loaded their arms with cloth grocery bags and supplies. Carl was already loaded with a large plastic sack emblazoned with a clothing store logo.

"Look at that!" She had stopped in mid step in front of the kitchen door.

"What?" Then he followed her gaze north, across the valley. "Oh!"

The last rays of sunlight kissed the snow caps in the far distance. Painted a bright orange pink by the atmosphere-filtered last light of day, the fuchsia mountain crest topped dark purple slopes below. She sighed heavily. "It's beautiful."

"Yes, and a far cry from what it will look like in 2150 — or 2220."

She turned to him; they were only inches apart. His eyes, dark on a good day, were impenetrable, black. This evening, they showed only a pinpoint of reflection from the distant mountains. "Carl, do you think we can make a difference?" She felt an awkward lump rising in her chest. An unexpected burst of emotion. "We've made a lot of progress in a six weeks, but…"

"Sweetheart, if there was ever anything that was worth the shot, worth going for, hell, worth changing our lives for. The difference between 2150 and 2220 is worth that shot."

She dropped her bags and leaned into him, surrounding his encumbered arms in her grasp. She swallowed the lump in her throat but couldn't stop the tear. She blotted it on his shirt and forced a deep breath to stop her urge to sob. Within a three-hour drive in any direction from that humble farmhouse, over eighty million people had perished, or would perish. She was almost glad she couldn't bear children. If her efforts failed, if Room for Tomorrow failed, any future generations in this corner of the planet would be laid waste, incinerated under the brilliant glow of multiple nuclear warheads. The little farmhouse itself would become a cloud of cinders blown downslope in the first pressure waves to blast into the future Neenach's glass and carbon fiber towers. She didn't want to think about it anymore.

She stooped and grabbed her bags. "Let's go."

~ ~ ~

"Hey, somebody. Can you give Carl a hand?" She called out. "He's got a lot more of a load than I do."

No answer.

"Grieg? Noel? Chu?" She called again, each name quieter, less sure.

"Shh. Maybe we came in on a sleep cycle." Carl set his oversized shopping bag full of clothing on the floor and moved to set two cloth grocery bags on the little dinette table.

Parker stepped around him, her heavy bags bare inches off the floor. Stepping around to the other side of the table, one of the bags tapped the cabinetry. Silently, a small panel opened and slid out, revealing some controls.

She only noticed because of the movement. A small drawer, or panel, with several knobs and instructions in four language boxes slid out. She had to bend to read the fine print; the drawer was about eighteen inches off the floor. She noticed, looking at it, that there were several fine horizontal lines in the cabinet work above the extended drawer. Indications, she thought, of more control works above them. She stood and exchanged a glance with Carl, who had just noticed the new little drawer.

"Odd place for a control panel." His voice was above a whisper, but not by much.

She said, in the same quiet tones, "I think there are other controls here." She pushed at one of the narrow rectangular indications of a drawer. First, a light click sounded from inside the cabinet; then it slid out, quiet as a cat. Again, each dial had labels in four languages. She leaned over again to look at the fine print. English, Chinese, maybe Japanese, never sure, and German. Yeah, German. She puffed a laugh. "Their labels are almost twice as long as the English." She bent over closer to read. "Looks like power setting controls for something. There's temperature readings, amps, kilowatts, a voltage gauge reading just over twenty-four volts."

She tapped a third drawer front and got the same result--a click of a release and another panel opened. The small books inside were emblazoned with a logo consisting of a 24-hour clock face. Words inscribed inside each book's circular clock logo read: *United Nations Historical Society-Time Service Field Guide, Energy Systems, Maintenance, and Emergency Procedures, Part 3, Recommendations for Time Local Excursions-1960–1980,* and *Part 4, Recommendations for Time Local Excursions-1940–1960.*

"Is there a part one or two?" Carl asked, leaning over his bags on the table.

Parker shifted the booklets around with her finger, feeling very much like a trespasser. "No, I don't see them. Maybe they weren't issued if they didn't plan on visiting after 1980."

"That figures." Carl said. "If the Last Day event was about fifty years before their time, about fifty years before our time would be the 1970s. This station seems to be set for a given distance back from its local time."

A bell tone sounded at the other end of the side room used as part kitchen, part laboratory, and the small shower sized decontamination chamber. Parker pressed each of the drawers and in sequence, each retreated back into its housing. She straightened and began to pull groceries, canned goods, and staple starches out of her bags. Chu entered first. They could see Noel's dark-haired form behind the translucent panel in the decon chamber.

She hoped she didn't sound guilty of trespass or worse. "Hi, we didn't expect to find the place empty. We, uh, brought you some food that doesn't taste like, wha'd you call it? Freeze dried sheep drops?"

Chu stood still, staring. She couldn't tell what his expression told. *Certainly not inscrutable. Embarrassed? No, why embarrassed? Is this what Chu looks like when he's pissed?* "Hope it's OK? Carl has some fresh fruit in his bags." Carl dutifully lifted a bunch of prime yellow bananas clear of one of his bags and displayed it.

Chu finally said, "Hope you closed de park road gate behin you. You can't leave it open. New visitors will be a problem."

"Not a problem, we've been taking the long way around the mountain." She looked aside at Carl for confirmation.

He nodded yes, then asked, "How long since you had a fresh banana?"

Chu looked up at the small skylight, considering. "Six or seven years. At least." His face again took on a look that blended confusion, questions, and some unreachable conclusion.

She beckoned him closer. "Come look, we've got clothes too?" She waved a hand toward the large white plastic bundle on the floor. "We could take you out, maybe go down the other side of the mountain sometime." She flashed one of her jury-winning smiles. "Sure can't take you out in tunics with built-in solar collectors, now can we?" To her relief, he took the few steps over to the table to look at the growing pile of food.

The decon chamber bell tone sounded again, and in moments Noel was standing at the door of the decon chamber. His face opened in gap-mouthed shock at the two unexpected visitors, and their pile of food. His head tilted in question, eyes narrowing. "Wasn't expecting you to come today."

Carl answered. "It's a Sunday, for us. The weather for a drive was unusually nice, so we took a road trip and thought we could make it here before dark. We almost did." Noel's look of unease was increasing. "And we found an easier way to get here instead of taking all the park roads across the mountain. We took the 5 north and then hooked back east to Neenach. Come, look at the goodies. Is Grieg coming too? We've got clothes for you all."

"Grieg's still in the valley—he's got the other cart wit 'im," Chu said, turning to Noel. "Come on mate, have a seat. Good vics here. Fill you tum with sumpin good gone down."

Noel's face was frozen; his thin lips held something back. He finally said, "After dark, you need to keep with the mountain road. It's too easy to find this place from the valley. If we can see the valley, the valley can see us. Sabe?"

"Uh, sure. OK." Carl held out hands in a no-contest gesture.

Parker was unsure if Noel's hard-to-please attitude was real or tough guy put on. But she made room for him to sit at the table.

Chu had been shuffling through the collected pile. "Wot all have you brought? We don' have room to stow all des victuals."

"Well," Parker began, "you said you'd be here longer than you'd planned because your team leader had died?"

"Right." Noel sat and looked at food that wasn't sixty years old or hadn't been extruded by the recycler's compiler. "We're down one crew here so we might have storage. What did you bring?"

In the next few minutes, the two visitors unloaded new clothing to replace the worn jumpsuits issued to the Time Service. Some of it was for simple comfort, and some would allow them to make forays out in the 21st without attracting attention. The last to appear, Carl had saved it for last, were serious-environment isolation jumpsuits. The barrage of new clothing ended to be replaced by the parcel-by-parcel explanation of 21st century foods and their packaging. Ignoring the jumpsuits, Chu grabbed at a purple athletic tee in the pile of clothing.

Noel relaxed and was fully enjoying the change in tension. A few minutes of genuine fun ensued with the opening and sharing of a bag of bakery-fresh chocolate chip cookies. His initial unease of finding the 21st visitors trespassing had passed. Noel was grinning at the names, common enough from his studies of the period, but here they were in his hands, spread out on the table. Fritos, Coke, Jiffy. "This is far gone, mates, cooked!"

They all heard, the outer door to the 23rd, and then the detox chamber cycling as the two residents of CalStation looked through the artifacts from their past. Grieg entered with, "Life is tough, but it's tougher if you're stupid."

Chu and Noel looked at each other, searching for the answer.

"John Wayne?" Carl offered.

"Buddha Balls, he got it."

"I was a fan, saw a lot of his movies when I was a kid on the streets. There was a fifty-cent movie house down the street."

Chu waved Grieg over. "Lookit!" He waved at the spread.

"Wassit?" Grieg picked up a can and read, "Campbell's Tomato Soup." I thought that was just art. There really was soup?" He began shuffling through a pile of folded clothing. He grabbed a shirt and got up. He went into the side sleeping chamber and returned wearing a fresh Lakers shirt with the team logo huge in the center of the purple shirt.

Smiling, Chu held up an odd flat black cube. "Got somtin for you in trade." When they looked at it more closely, they saw that it was decorated in a fractal pattern that resembled tree branches etched in silver. The blue-black tint gave its true purpose away.

Carl asked, "Photovoltaic?" Chu handed it across, and Carl turned it over a few times in his fingers, examining it. Carl asked, "How, um, why do you do solar PV in a cube?"

"We don't; it's a PV wrapper. Paper thin, bonded to Mylar, and shrink-wrapped to the cube." Chu described the simple functionality of the object. The little cube was powered by shrink-wrapped PV surfaces on five sides, charged by room light, power stored in a laminar carbon battery base. It held 2,500 terabytes of data storage. The sixth side was the contact side for data transfer to their consoles. The cube could hold more information than the world had produced before 1950. But the information Chu was especially interested in giving to the two visitors concerned the manufacturing technologies that created the first layered-carbon batteries. That technology, if "discovered" a few generations earlier, could almost certainly pave the way for large scale solar in the American West, the Gobi, and North Africa. That might avoid at least two of the Middle East wars of the 21st century and preserve the valuable oil reserves for much more necessary plastic and composite materials.

Chu held the 4cm cube up, absently twirling it on opposite corners. "Maybe, just maybe, we could hold off the Second Jihad or the Cape Town massacre, and who knows, mebbe even the Last Day."

He held it up almost at eye level, so he was looking past it at Carl. "It's for you." He reached out again and lowered his palm for Carl, who took it and rolled it over in his fingers, examining it closely.

"Fantastic! How much did you say it holds?"

"Far more than you'll need; 2,500 terabytes."

Carl blew a silent whistle.

Parker's mouth dropped open. "You're giving that to us?"

"Parker, Carl, I need to figure out how to get information *from* this to 21st century tech." He waved a hand in a helpless circle. "We can't share data with your hardware. It's incompatible IO with our tech. I'll have to think about this."

Noel leaned away from Chu; his face torn in disbelief. "The hell, Chu. We hadn't decided yet." He got up from the table, growing fury barely suppressed. "You still need clearance from Wellington!"

"Noel, please, calm yoself." Chu rose too, his body language a picture of supplication.

"Calm myself?" Then in higher tones, incredulity coloring the repetition, "Calm myself? Buddha's balls on a biscuit. You shyte!" Noel grabbed at a breather mask, stepped into the decon chamber, and cycled it to let himself out into the toxic soup outside the back door.

~ ~ ~

The aching quiet left after the portal closed with a soft sizzling pop remained for an uncomfortable too many seconds. Chu got up and went into the head and shut the door. Grieg broke the silence. He waved to Parker who appeared to be near shock from witnessing the emotional breakdown between Chu and Noel. "Com'ear. We showed you the worst side of the future last time we showed you anything. Lemme share some things I put together for you." He waved haptic enhanced fingertips at the holo screen and called up a vid.

Parker and Carl moved closer to Grieg's workstation. His viewscreen took shape and began to display scenes that could be recognized as Los Angeles. City Hall's needle was barely visible among the newer high-rise structures. But it was not a view of the smog filled 21st, nor a vista of ruin of the post-apocalyptic 23rd. The aerial view flew through gleaming bronzed, silvered, and gilded high-rise structures. The viewpoint ducked under pedestrian crossing bridges at impossible altitudes, then dipped along boulevards filled with linked cars. A cutaway scene showed an ovoid ground car disconnecting from a linked set of similar, but not identical, vehicles. The newly isolated car entered an exterior elevator pad and rose rapidly to a bay two-thirds of the way up a hundred story building. A man in gray formal business attire, or what seemed to be formal business, exited the car and stepped onto an elevated platform. The car moved into a small cubicle just before its elevator zoomed back to ground level.

The scene changed to an interior showing a birthday party for a young girl of around ten. She and some friends were celebrating with cake and presents much as Parker would have imagined. Giggles and laughter as a father took video.

The next setting showed the same family as backs of heads and a profile of the ten-year-old's slightly older brother as an air car, gaining altitude, flew between high jungle-covered peaks. The next shot showed the family outing gathering for an awkward group photo with the stone architecture of Machu Picchu in the background.

The scenes continued with some bucolic family game time, a trip to an unknown beach with the children slightly older, and another man as father figure. Less obvious fun on record, but still a family outing recorded for posterity.

The series of images closed to a black dissolve and a credits page of a professional events recording service, media links, and a five-second commercial spot. Grieg spoke into the quiet. "Some of them, at least some of the middle class who could have their family's events recorded like this, had it good. We have to hope that enough of the rest had it good enough that it will be worth saving." He winked at Parker, threw a quick nod to Carl, and breaking an impish twist to his lips, said, "I wish you luck!"

11

The new offices of Room for Tomorrow were anything but a simple storefront. They had moved into the third and fourth floor of an older six-story building just outside of downtown Anaheim. With an option to pick up the top two floors in a few months and the bottom two the following year, they had growth potential in the building and plans to exercise it. Megumi, Parker's capable and trusted assistant, was quick to adapt and had taken charge of the mundane and the difficult.

Parker was pleased that she had made her Room for Tomorrow's new office manager. Meg had taken on with zeal the role of chief cheerleader and woman in charge behind the scenes of the new environmental advocacy group. She'd admitted after the transition into her new position at RFT that she hadn't been happy dealing with most of the couples that came through the divorce, custody, and child support gauntlet that was the California legal system. She'd been on the verge of quitting.

In the first month, cubicle dividers and furniture had been installed, and telephones — banks of them — were operational. New staffers were making and taking calls immediately. Social media pods, in pillow-cushioned quiet rooms, emailed, texted, Facebooked, and Snapchatted in the world's most affluent languages. Potential cell leaders were located across the US and around the world, drawn from membership lists requested from allied interest groups who were given significant donations in exchange. In two months, the call center had filled with personnel hired on the dual criteria of ideology and social media skills.

New hires were required to view key documentaries on ecological damage around the world, climate change, and industrial disasters. They were well-compensated for their time off the phones and became that much more dedicated to the cause with their increasing awareness of 'the problem' as it had become loosely defined – impending global catastrophe. Following weeks of furious activity, Parker and Carl were ready for the first Board of Directors meeting. The Board was not hard to find in the Los Angeles area, but they had reached out to northern California, Washington State, and even Chicago for key individuals with ties to the green industries.

Carl watched as Meg entered the new office's boardroom, illuminating the gathering with a broad smile and warm-from-the-printer agenda packets in hand. He matched her grin and saw Parker at the far end of the table returning it. Many of the newly selected board members were involved in introductory conversations or catching up with acquaintances made in other circumstances of the environmental movement. He rapped knuckles on the polished oak table. "Madam President, I believe we are all here."

He scanned the six faces along each side of the conference table and then caught the sparkle in Parker's eyes. She nodded for him to continue. "Ladies and gentlemen, we've been busy these last few months. In the server room, ten new server racks are connected to the fiber loop in the street with links I supervised myself." He coughed into his fist and smiled. "Nice piece of work if I do say so. The last bits of hardware were special ordered on a proprietary design and secured so that Joanie and I have the only access." He pointedly looked to the attractive-in-a-nerdish-way twenty-eight-year-old Joan Brayton who had impressed him far beyond the field of applicants interviewed. "I'm proud to say we have enlisted two new employees to keep a strict and personal eye on cybersecurity. Our platform absolutely *has* to be impregnable." At his nod, she stood, smiling. "Her assistant, I hear, is in the basement with wiring pliers in hand." Smiles and nods all around.

"Our public profile has grown!" He spread his hands wide, palms up, and then with a sense for the theatrical moment clenched his fists. "No breaches can be allowed that would let any of the many SuperPAC-funded science-deniers, and creation-believing, fundamentalist conservatives tap into our agenda, our phone lists, or our donors." He scanned their intent faces. No one was checking a phone or in any way distracted. He had them.

"And now, I'd like to hand off your official welcome to my inspiration, your director, and the woman who wears the pants around here, Parker Parrish." Smiles, applause, and quiet laughter spread around the table. Parker was, as usual, in blue slacks.

She quieted the room with outstretched arms, palms down. Her joy in the moment was impossible to hide "Our battle lines are being drawn across the environmental protection spectrum, and in Carl's tech department, they are being drawn through a new website based on critical evidence and first line research. We have the solid evidence that the world is going to hell faster than anyone in political power wants to admit." Smiles and nods of agreement answered.

"I want you all to check out the website, look at the areas of interest you're most familiar with. If you can find newer studies or commentary, please suggest them. We'll add them to those tabs." She picked up her own set of the identical piles of paper placed at each seat. "Each of you has the complete site map and contents printed out. You'll find a tab labeled 'Unbridled Greed.' We also want to document for the world and, in particular, to our donors, that greed incorporated has a hand in a lot of unnecessary human misery, habitat destruction, and species death.

"If you look around, actually look hard into all the trouble spots in the world where there is conflict, you'll find that many of those are based on extraction-based competition. Many of the rest are based on sectarian religious conflict and on long-term political struggle. When you dig deeper, most are again based on scarcity of resources. In World War II, the Japanese took over the Pacific Islands and Southeast Asia because they needed oil and they needed rice to feed an army. A few years ago, the Russians re-annexed the Eastern Ukraine over oil resources and ports. Same with the Latvian port facilities in Riga." She shrugged, "OK, so the Latvians have political autonomy, but it's accepted that Latvia scratches when Moscow itches.

"Those are only a few examples of superpower bullying. But they are about oil. Other conflicts are over scarcity of resources, amplified by boredom and exacerbated by lack of secular education. There's a laundry list of conflicts where unemployment above thirty percent is endemic. Most of these arise because birth control is suppressed for cultural reasons, and out of control population growth exceeds the carrying capacity of food and basic service delivery systems. Religious differences, and seemingly

105

minute philosophical differences in dogma, further fuel the conflicts.

She began to tick off a list on her fingertips. "Iraq, Afghanistan, the Palestine, Syria, Ethiopia, Sudan, Libya, Donbass, the Burundi, Mali, the Niger Delta." She stopped and looked up at the gathering. She'd run out of fingers. "Humans have a seemingly unfillable capacity to kill each other off if they can only call God in on their side to justify their actions." Parker stared off beyond the roughly formed rock wall that would soon be a water feature, looking inward for her next thoughts. "Underlying all of these conflicts is poverty. Where there is adequate nutrition, you see a lot less conflict. Europe, North America, Australia, and New Zealand." Her mind raced to the inescapable but impossible to share knowledge gained in the little farmhouse portal. "Freaking New Zealand!" Her voice was rising. "They have it going on, don't they? They share the world's economy and spend but a pittance on a defense force." She shook his head, thinking of the three Time Service guys in the farmhouse a few hours' drive north.

"If I could tell you —" She paused for a moment, considering. *Bugger it.* "If I could tell you that I've seen the future and that the future is a horror show, you'd smirk and think I'm crazy, or delusional — but you would have to admit to yourselves that none of the indicators are good. None! The world *is* going to hell."

Shit, I'm ranting. Parker blew out the frustration that was framing the next sentence. After a beat, she finished.

"So, I'll stop yelling at you. I just want to say that we, you all, all of us here at Room for Tomorrow, are taking a first big step toward actually doing something beyond lobbying and public education. Not that we won't be doing those, but through our fundraising steps, from Kickstarter and beyond to direct investment venture capital funding, we are going to be making a difference."

Parker paused, lips pursed against any additional preaching, and let her mouth relax into a smile. She opened her arms expansively, inclusive of the attendees, their new start, and in her mind, saving the world from itself. "So, a big welcome to the Board of Directors of Room for Tomorrow!" She took in each of the faces that were fixed on hers in wonder. "Thank you for joining our board. We put a lot of thinking and work into selecting who would be sitting in your chairs. Each of you has an area of expertise that informs energy, or climate, or weather patterns, education, health, poverty, food supplies, reforestation, population control, seed banks, transportation networks, and—" she paused for breath again, "—did I get everybody?" There were some chuckles around the table.

"Oh yeah, media, and research, and public-private finance!" She took a mock bow. "We have a lot of important work to do— education which includes debunking fake facts, complex systems analysis, funding appropriate technologies, doing assorted good deeds, researching, and then applying solutions!" She'd run out of breath. She picked up her agenda packet and poked her left index finger at it. A glance around the table told her she'd affected most of them. "We've got a lot to do. You'll find committee tasks, data references, and personnel contacts for each of your areas of responsibility. So, let's make a difference; and finally, thank you!" She dipped again in a final gesture of an exaggerated stage bow.

~ ~ ~

Carl clapped first in moderate measured beats; as the others joined, the tempo picked up. It seemed they all stood as one, and Carl leaned to shake the few closest hands. Parker's cheek betrayed a single line of moisture, but her eyes weren't the only damp eyes in the room. The moment was over quickly in a self-aware silence. Parker moved first to sit down, and the other eleven followed. It was Parker's invitation to join the board of directors that had brought them all to this moment. Now they faced her. She allowed everyone to compose themselves while she took a drink from her tumbler of tea and re-shuffled her paper stack back into order.

She began slowly. "Every two years, the American electorate proves it can be swayed by micro interest groups to vote on single issues instead of understanding the big picture. It can also be cajoled into thinking that outside groups and elements are at fault for all of their woes. The 2016 elections in total," she allowed a terse puff of a laugh, "were disastrous for the environmental movement. We had science and education committee chairs in Congress who were either completely demented or who actually believe that the earth was created 6,000 years ago, despite historical, scientific, archaeological, geological, and astronomical evidence to the contrary.

"There is no way any politician is going to bring steel manufacturing back to Pittsburgh or make West Virginia coal clean enough to be a viable portfolio option for safe energy production. What they wanted to do was remove from the books the Clean Air Act, the Clean Water Act, and further emasculate the EPA. Room for Tomorrow's public image is *our* new stage for raising consciousness. Our war chest is growing. Funded by thousands of individuals who voted with their credit cards. So let's start. As this is our first meeting, I'd like each of you to do a brief introduction and share some thoughts. Think about the linkages as we go around the table and see who among us is most likely to help each other."

She didn't mention that their bank account was overwhelmingly funded by first the Abadons; then Chained Industries LLC, the oil conglomerate in Houston; Mexico's state owned petrochem, PeMex; a French Chemical conglomerate; and even Rosneft and Khimprom, Russia's largest oil and chemical producers respectively. Deposits and transfers from those companies were always going to be behind the wall. Behind Carl's hidden doors.

Room for Tomorrow's first board of directors meeting adjourned four hours later; inspired, and informed in great detail of the organization's planned public face. Parker was pleased. Their new environmental start-up was well funded and could grow into a major new world player. Some of the board, she thought, were just happy to have a new plum entry on their CVs. Most would be valuable assets through their connections in industry and government.

Parker exited the elevator, dressed in her new office uniform — navy slacks, sandals, and a comfortable white blouse. She'd sworn off suits. Carl met her as he returned from a quick side trip to his office suite. The complex of rooms held his office and an assistant, but most of the rooms were filled with rack storage and servers.

"Hey, good looking."

She looked embarrassed. "I went overboard, sorry." Her sheepish shrug and boyish grin were disarming but unnecessary.

"I don't know that you did any damage. I actually thought you showed restraint, considering."

She looked around, gauging earshot. "I was proud of you too. Let's go." She linked him by the arm and kept moving. He had to shuffle to keep up. "I have a side project I want us to consider," she continued, her voice a tribute to her heightened enthusiasm. Waving her free arm at their new offices, she said, "That's not the royal us — that's you and me."

"OK, what and where?"

She guided him into the stairwell, and as their steps filled the echo chamber with the taps of leather on concrete, she continued. "First, have you ever read a book called *The Monkey Wrench Gang*?"

"No. I can't say — " He let his silence ask the follow up.

"Forty or so years ago, Edward Abbey wrote *The Monkey Wrench Gang* in the hey-day of the hippie movement. Someone left it at a beach house I rented at least twenty years ago, but it stuck with me. It was old even then."

"That's impressive for a book to have that much staying power."

"If I remember correctly, either it or its sequel, begins with a turtle cruising through the desert scrub experiencing a rumbling, earth-shaking vibration that it's never felt before. Its reptilian brain isn't up to figuring out there's danger attached to it until it becomes buried and then scooped along in front of a bulldozer that's cutting up pristine desert for a water guzzling subdivision. Pretty powerful."

"OK, so what about the book?"

"The monkey wrench gang was a group of environmental activists who were into industrial sabotage to construction equipment and the like, trying to slow down or discourage wasteful sprawl development that was encouraging people to live forty miles or more out in the desert, simply because the land was cheap. The book tried to publicize that economic fallacy of desert sprawl and much more." She slowed; her enthusiasm had spent her last breath of air in explanation.

They stopped at the mid floor landing. "Well, it sounds like I'd have remembered it if I'd read it." He smirked an "I'm sorry."

"Oh, don't worry, it was before my time and yours, and I only found it as a worn paperback in a rental house. I want to fund a new edition of it and the sequel, *Hayduke Lives*." His eyebrows lifted. "And I want to donate them to libraries across the country, specifically as a gift from Room for Tomorrow. There are other more current titles too."

"That would do what exactly?" He seemed unconvinced.

She persisted. "It would get the books and the idea of environmental activism back on the front burner and save us from having to do some front work with a new writer. It's a ready-made challenge to activists to go do something about whatever is pissing them off." She could see that she was making some points. "There are a lot of other books out there too that need promotion. Maybe we could start a trend called 'Green Shelf Books' or something."

"How does donating a few thousand copies of a book to libraries make waves?"

"Because I'll get airtime on CNN and PBS and wherever I can. *The Today Show*, Jimmy Kimmel, Jimmy Fallon." She beamed up at him as they stopped walking. "And that will lead to my Phase Two project." She could tell he'd follow her to the slums of Caracas if she'd only ask. She could see his dark eyes dancing across her face, glistening with emotion.

"Ah, all right, let's do it; after you convince me of Phase Two."

She lifted up on her toes and gave him a peck on the lips. "Then come into my office." They emerged from the stairwell onto the executive floor, and she led him into an office that had a "not moved into yet" look. Boxes sat unopened in front of her shelving. She had managed to get her JD diploma on the wall. An acrylic "Young Lawyer of the Year" award from a decade back sat alone on a shelf. Her desk was littered with sorted piles that indicated a nascent personal filing system was beginning to fail.

As she feared, his first reaction was, "Christ almighty, Park!" He waved a hand around the paper storm. "You need Meg to come in here and save you. I thought you were bringing her over as your personal assistant."

"Not anymore; she's too valuable running the office. I don't know how to do that, and you certainly don't."

"But—"

"—there's almost 200 of us now. I can't and don't want to manage that. Meg's been great; and she's great at it. Me, I want to hold the spear, drive the car, write the outlines." She held her hands up in mock surrender. "I gave it up. I was just writing up the job description last week and realized Meg was it. She's the

Chief Operations Officer." She opened up her right hand toward Carl. "Unless you want to do it."

He backed up a step. Horror framing his face in a mask. It softened into a grin. "You know my strengths, and they aren't that. I'm more of your behind the scenes, back office kind of guy. I have to be; and I'm glad, too, that you are handing off general operations to Megumi, good move." He stopped, remembering why he'd come to her office.

He puffed out a last breath and changed channels. "OK, what's Phase Two?"

She pulled a colored map of Florida off the printer that indicated the state's outline with several red or blue lines running into the state and down through the peninsula's interior. At first it looked like an interstate system map, but he realized that it couldn't be. They were not in the right places and had strange letters identifying them instead of I-10, I-75 etc. "Those are the gas pipelines that feed the state of Florida," she said. "The last governor was so damned proud of his pro-business agenda that he either ignored, backed away from, or defunded most of the existing environmental programs in the state, including his own environmental agency."

He nodded in understanding. "The jury still out on the new guy?"

She shrugged, "We'll see," then poked an index finger at the lines on the map passing into the state from Alabama. "But it doesn't look promising. The state gets more than eighty percent of its energy from those pipelines. Some are gasoline, some are natural gas. Generator stations across the state get their power from burning gas that's pumped into the state through Georgia from the north here and all the way from Texas into the panhandle here."

She traced her fingers along the set of blue lines that ran generally eastward from the petroleum fields he knew extended from West Texas to Louisiana.

He asked, "What are the gray lines? Oh, roads?"

"Yes, and they cross the pipelines in some very isolated locations." This time her eyes danced with mischief.

~ ~ ~

CalStation always seemed crowded when the time station's crew was joined by their visitors from the 21st. With Grieg in the valley, it was a little easier to move around. Chu wore a new polo shirt emblazoned with yacht club logo, but Noel still had on one of his original tunics with charger panels on the yoke.

Parker watched as Carl pulled several bits of techno junk from his backpack. He looked across the table at Chu. "This might help our problem." He set a black object on the table. The small rectangle about half the size of a placemat opened up to reveal a keyboard and screen. He reached deeper into the backpack and pulled out additional equipment.

Noel sat down beside Chu as Carl arranged and cobbled together the bits of techno junk. "This," Carl said, waving over the pile, "might be a way to get information from your data cubes straight to our memory drives in the 21st." He began plugging jumper wires into a small pink-and-yellow component that looked like an elaborate child's toy or a pet's food dish. At the other end of the wires, a hand-soldered green circuit board had a simple knurled knob at one end but no dial or scale.

Carl finished with the plugs and connections, looked up, and smiled. "There! I think that's got it. Do you have a data cube handy?" He had been typing intermittently during his connections, and the little keyboard thing emitted a musical chord as the screen's display changed, offering a command prompt in a black rectangle.

Chu leaned across the table, peering at the unfolded black device. "Is that a real laptop?"

Carl swallowed a laugh and said, "Yes, genuine 21st century article. Not the fastest thing on the market, but it's got a terabyte of memory, and we've brought some external drives that can triple that."

Noel whistled, mocking an impressed tone. "Hear 'em, Chu. A whole terabyte!"

Chu answered the whistle. "Well, le's put a fork in me arse and turn me over, I tink I'm done."

Carl huffed. "Listen, guys, this could help get whatever choice technobits you can or are willing to share with us into our data storage. Maybe we can speed up the tech curve and solve some of the problems that are going to cook your grandparents."

Noel straightened his spine. Why did these visitors from the past think "the problems" that instigated the Last Day could be solved with a technology injection? He said, in cool even tones, "Carl, don't get me wrong, I think what you two have done, are doing, will save a lot of suffering in your time and for a few generations to come." He felt Chu's hands pressing on his calf muscles, tightening. "Chu? Why are we going to give them access to technology they can't even machine up to in the foreseeable future?" He gestured to the pile of equipment Carl was assembling. "Is this your doing?" The hands on his calf relaxed their grip and pulled away.

Chu said, "Noel, we talk about dis. Please give des blokes a chance to make 'em a difference."

A soft ping tone began to come from the communications console. The mounting tension abated, for the moment. The two men turned in unison to view one of the time station's screens. The holo screen had formed unbidden. As its imaging plane took form, the screen resolved to a simple digital display; a single green dot blinked beside a frequency indicator. It stopped blinking almost as soon as the image solidified, but the frequency indicator remained. Noel left Chu's side, pulled a retractable seat from the counter and sat in front of the console. "Whatever dat was, it's gone except for dat." He pointed at a circular display with a sine curve frequency wave form sliding by.

Noel tapped at a few controls but was unable to track the source of the unexpected signal. "Well, bollux!" He had just turned from the display when the warning ping sounded again. He turned to the console to see a green indicator go dark again. "What the..."

"I think it's me. This thing." Carl toyed with the single control knob on his gadget. The green indicator responded each time a turn of the frequency modulator hit a sweet spot. "Yeah, it's this tuner." He watched as the sine curve tightened or spread out in response to his fingers on the dial.

Noel stared at the young computer tech from two centuries back with new respect and some caution. "What exactly is that?"

"This," he said, pointing to the colorful could-be food warmer, "is a kid's toy. They place game pieces on it, and that triggers different game play based on which action figures they put on it. The child's game avatar is triggered by a code magnetically embedded in the game piece. I was hoping this rig would be able to read data cubes or at least download the data stream for decoding later."

Noel looked up at Chu; his face reddened, his carotids swelling. A single vein bulged from the center of his forehead. "Chu, the hell?"

"Easy, easy, mate."

"Don't mate me. I never agreed that we would wholesale our tech back to the 21st." He attempted to stand, to face Chu, but couldn't rise against the renewed pressure of Chu's hands on his shoulders. Humiliation fought with betrayal. And the two visitors, Carl and Parker, could only look at each other with raised eyebrows. Noel, in barely suppressed fury, tried to maintain civility. "Carl, when did Chu tell you to get this gear together?"

"He didn't," Carl and Parker answered more or less in unison. Their apologies tumbled out in overlapping bits:

"… only trying to help—"

"…didn't mean at all to upset—"

"so sorry"

"thought that it might be understood, you—"

"… avoid a holocaust."

They heard the sound of the decon chamber cycling again. Carl noted that upper translucent panel was opaque enough for decency when used as an undressing chamber, and transparent enough to see who was coming through. He could easily see that Grieg was coming in from his mission down valley.

Noel raised a hand to restrain Carl's enthusiasm. "OK, I get it." He raised his head slightly, indicating Chu who was still standing immediately behind him. The pressure from the moving fingers on his shoulders seemed to be trying to become a massage. He brushed at, then slapped at, Chu's fingers. Carl was still across from Noel and became his target. "This bloke's put the idea in your head that we're all partnered up now in saving the future. Hasn't he?" Carl looked up to Parker for help. She looked back; thin-lipped, wide-eyed.

Noel shook his head in frustration. Took a breath. "Listen, you two are all right, and your kibble from the outside certainly beats the service rations we brought. And thanks for the clothes. It might be nice to get out into your world. Might help our accents, and certainly his." A flick of his head indicated Chu. He felt his heart rate slowing. "We were wondering if we'd have to break into the rations left here by the researchers from before. Didn't want to. In our time, it's over sixty years old."

Parker tried again. "Noel, we're sorry. Carl's talents aren't just software and security. He talked to me about it and—"

Carl finished, "...it seemed like a good idea to be able to build a bridge to your tech from ours. We could use some breakthroughs. I've heard you guys talking about carbon fiber conductors. And laminar carbon batteries. Hell, we're still using carbon-rich rubber as an insulator, not a conductor."

Noel shrugged off Chu's hands that had gently landed back on his shoulders and stretched his neck muscles, trying to throw off too, the emotional connection Chu was trying to maintain.

Grieg finished in decon and stepped closer to the table. "What was that beeping about?"

No one answered for too long as practically everyone exchanged looks with everyone else. Carl answered. He could see the frequency displayed; 2.40GHz. "It's Wi-Fi." His boyish grin began to spread wide. "I think I've accidentally found a way to directly tie your electronics to our internet." He addressed Noel. "If you'll let me work with Grieg on it, I think we may be able to leave this laptop with you so you can access our century's communications network. I'll leave you a couple of phones to use as a data source for the laptop. You can check the weather. Check the news, see what's going on outside."

Noel was only partially mollified. "So, we can see what was going on 194 years ago?" He snorted his disdain. "I can also look it up. It's history; it happened."

Parker chimed in. "We could also bring you some of our available surveillance equipment. If a wire can pass a signal through your, uhm — that force field—you could set up intruder sensors. You could set up approach cameras. You'd know when to shut down the field to save energy and when you had to turn it on to avoid detection. Once we set you up with a tie to our tech, we could even set up a link."

Noel continued to listen but heard little more: throbbing in his ears matched the thumping of his chest. He was more than angry at Chu. He didn't believe the "accidental discovery," the "we only wanted to help." He sensed that Chu had moved away not only physically but emotionally. A quick glance behind showed Chu standing closer to Grieg—far enough away. He stood, made a show of straightening his tunic, as if the material could wrinkle, and moved toward the portal. He glanced up at the dual time clock to check local time in the 21st. "The moon'll be up soon; I'm gone for a walk." He grabbed his mask and a sidearm and was gone.

12

Noel worked quietly at his workstation. His fingertips flew across bare spaces, weaving intricate patterns that he and his terminal had fine-tuned for speed, accuracy, and his hand's individual geometry. The implanted haptic fingertip interface was faster than the ancient keypads in use in the 21st. He had tried the laptop that Carl had provided to get local internet services via something called a sat line. His fingers had kept slipping away from what Carl called 'home' on the keypad.

Now, with the primitive laptop's internet connection linked into the CalBase time unit's intelligence system, he had access to both his own data systems and that of the early world wide web. Despite his early resistance to the outside help, he was amazed that the linkup had worked and thankful, because now they could get local weather alerts without poking their heads out the farmhouse windows.

The weather outside was fine — out the back door to the 21st, sixty-five percent humidity, mid-eighties, and typically clear central valley skies. If he went out the front door, he'd find hellish heat, toxic air, and background radiation levels that would eventually kill them all. On any given day there was a good chance that a dust storm would descend from the northern valley's expanses of bone-dry valley floor.

"Chu?"

"Yes." The reply was muffled by bread and chewing noises. He heard a slurp, then a much clearer, "Yes, what is it? Have you found something?"

"This cube has a billion somethings on it, but I'm struggling to find anything pointing to their latest fuel cell research. A lot of it was encrypted."

Another slurp from the pull-out kitchen unit. "You'd tink de containment specs for cold fusion would be something dey kep in any library."

Noel gave a shrug, invisible from around the partition. "Seems corporate lawyers insisted that it be shrouded. It's secured behind firewalls. I'm seeing the traces, and I know where it must be. Grieg will have to take a shot at it. I can't get through the encryption."

Chu walked up behind him, circling the back of Noel's neck in strong fingers. Chu's thumbs began to knead at the pronounced spinal ridge. Although Noel would never have been labeled a chunk, a pejorative term for overweight on NZ's North Island, he was both small in stature and not too many kilos heavier than Chu's thinner frame. The argument of the previous week had dimmed but not quite extinguished the flush of excitement he usually got at Chu's touch.

"Uhm, that's good." Noel relaxed his shoulders and let the massaging thumbs destress his neck muscles. "I'll give you fifteen minutes to stop that."

"Can't take that long; got something we need to discuss." He let his fingers rest on Noel's shoulders. "NZ Base messaged from the last flyover."

One of the few still working high-orbit telecom satellites provided momentary links between the Time Service's headquarters in Wellington and time stations that were known to exist. The Japanese unit in NagoyaBase, and CalBase were the only two known to be working and useful. AfricaBase in Johannesburg was too hot to use, and both DCBase in Arlington and the EuroBases in London and near Stuttgart were obliterated. Noel straightened up and turned a quarter-turn toward Chu.

"We get sailing orders?" Noel inclined his head upward to hear better.

"Nah, not even close."

"Crikey. If it weren't for the weather here in the 21st, I'd really like to rotate home." He waited. Chu was slow in following up. Noel turned a quarter turn toward Chu.

"Dere isn't gone be a rotation. NZ said to stay put, keep working. Dey can'na get anodder sub scheduled to retrieve us for another six months."

Noel erupted out of his chair and turned on Chu. He had to look up to meet his partner's gaze, but his anger was unmitigated by his size. "Six months! What the BLEEDING HELL? How much longer do they want us here for? My dad's gone now, and mom has leukemia. My sister has had another kid while we've been gone, and I'm half a world away, poring through ancient archives." He shivered in frustration. "Ahh! And we're running low on mess kits." Noel wanted to hit something, so he lashed out at Chu instead. "And Jesus, Chu, work on your fucking English. If we're going to go out in the 21st, you don't want to draw attention by sounding like a North Island gutter snipe."

"Calm yo self down, Noel." Chu had a look when he was in placating mode, and he had it now. It wasn't helping Noel's frustration. He continued his attempt at conciliation; speaking slowly now, apparently working at his fricatives with more care. "We now have frienly local contact. The food they brought a while back is pretty damn good — better than rations. The dew precipitator, she workin' fine, and you know you like our walks outside at night." Chu had dropped a reference to their interludes in the soft grass behind the farmhouse. Chu tried a subtle reach for Noel's hand but wasn't quick enough.

Chu then tried to put his hand on Noel's shoulder.

Noel practically spat. "Fuck you manno, and I do not mean that literally." He slapped Chu's hand away and turned to get to the satellite link. "I'm going to call Hickock and make an appeal. This is bloody rot!"

Chu stood in front of Noel, blocking any passage in the narrow, confined passageway.

"Chu! Please. Get out of my way! What's the matter with you, Solomon? Let me at least plead my case, I'm tired of this bleeding exercise in futility."

"You signed up for dis, Noel," Chu huffed. "You 'member looking at you contract? Seeing the chunk of slice of life you was signing away. Twelve months per tour; min-i-mum." They both understood that the twelve months would include the trek back over the mountain and a six-week journey in an antique sub back to NZ. "Twelve BLEEDING months!"

"RIGHT!" Noel shouted back at him. "I signed up for twelve, not more, twelve!"

Chu paused, not moving out of the way, and persisted. "Our mission here is more important than your sister's kids. The mission itself, it's important. Our next trip back, we'll be bringing the tech for haptic inserts and engineered cornea replacements. That entire file on genome-spliced cures for thirty or so cancers, coordinates for lodes of undepleted rare earth deposits, the stuff you working on now. Most of our CF reactors are running on empty." He pointed at the floor. "The one they've got in this station? Shyte, we'd need to bring that sucker back with us if we can't find the specs."

"That would kill this station for the next team."

"And, tink about it, Noel. What if dose two from de 21st can shake da can, make some noise, be heard? If day, no, if *we* are successful, our families have a future."

Privately, Noel didn't want to think about or, at this moment, talk about the new thoughts that had been creeping into his sleep. That if Parker and Carl were enlisted, if they could actually help by changing the inexorable path of history—maybe simply nudge it— then perhaps all those years into the time stream later, all those billions wouldn't have to be incinerated, poisoned, or murdered. So many changes downstream, mom and dad probably wouldn't meet, would we all simply cease? But here was Solomon, pushing against him, trying to block access to the sat link. He wanted to call Wellington himself. He resented Chu's assumption of the leadership role. He took a half step forward, resolved to see if Hickock had OK'd Chu's new plan.

"Solomon fucking Chu. Let me pass!"

"Noel, listen to me, please." Chu was desperate to keep his alternative alive. He tried once again to calm Noel down. "Noel. Mate. Sweetheart." Chu appealed now to the emotional depth of their relationship even while ravaging its trust. "The message *was* from Hickock."

Noel's eyes narrowed; he'd sensed something false. "Let me look at the goddamned message!" He attempted to push around the taller man only to be thwarted by a simple short sidestep. "Dammit! Let me through." He ducked to get leverage and pressed against a countertop.

Chu's fingers grappled Noel, first by the shirt and then, as the strain of their struggle increased, one slipped up to Noel's throat. Noel looked up disbelieving.

~ ~ ~

Chu now looked down at his friend, companion, and lover in real anger. Real passion, the passion of commitment, and his fingers were tightening around Noel's throat. With some horror, he realized what he was doing as Noel began to choke. When he finally released his grip, Noel coughed through reddened cheeks and again tried to slip by. The maneuver didn't work. This time Chu stepped back a half-step while shoving forward. Noel, caught off balance, lost his footing and staggered off balance. He fell to the side, hitting first the edge of the counter with head and shoulder, then down to the floor of the passageway. Noel moaned twice, shuddered, and lay still.

Silence. After the few minutes of violence, the silence in the station was sudden and sobering. A drip from the mini-shower and Chu's hard breathing were the only sounds. One of Noel's hands moved from hip to belly; one corner of his mouth ticked into a half grimace and relaxed. His body bent slightly, as if it wanted to curl but had forgotten the moves. The hand on his belly began to quiver, no, just the fingers, twitching as if tapping out a final message. Chu stared dumbstruck by what he had done, by what had just happened. Noel's fingers stopped twitching. The small pool of blood that grew slowly from behind his head made no noise. At Noel's workstation, an indicator piped a thin even tone in indication that it had lost brain wave connection with Noel Antonides.

"Holy bleeding Christ!" Heart pounding, breathless, Chu dropped to his knee and felt behind Noel's head, found the hot bloody gash behind Noel's ear. He gasped as his fingers came away sickeningly red and dripping. His high keening wail now filled the time chamber as Chu collapsed on his haunches. Realization sank in as his loving touch to his mate's neck revealed no pulse. He leaned closer, hoping for an eye flutter, a breath, a sigh. He drew closer still, holding the thin flesh of his eyelids over Noel's mouth, praying for a breath. As his own lungs emptied, unwilling to refill, his thumb again went to Noel's carotid, again searching for any sign of life.

Nothing! How can that be? Why wouldn't you simply listen? Wait a bit? See if any change at all occurred in the timeline? Now, look at you. Look at us. Oh bleedin' 'ell. On his knees now, some mechanical instinct drew his hand up to close Noel's eyelids. You are lovely, my sweet. A tear ran to the tip of Chu's nose, swelled, and fell onto Noel's cheek. Chu leaned over, kissed the tear; and then as tenderly as in life, kissed his partner one last time on cooling lips.

13

Lying in bed in the dim green glow of her alarm clock, the vestiges of a dream lingered as Parker slowly rose to consciousness. She grasped at wisps of scenery as it faded. She had been playing with high school friends at a local swimming hole, one of the hundreds of karst sinkholes in the woods surrounding her region of North Florida. Senior Skip Day. The last day of real school before the mundane half-day rituals that for some reason occupied most of the last week of why-are-we-still-here in high school days. The dreams had played over and over two scenes with subtle memory loss variations. The first, grabbing rope swings by their granny knots and swinging out over the cool green waters. The crowd of light beer-fueled friends would then shout jump or dive at the end of the swing, and you somehow contorted to try whichever shouts had been the loudest. Belly flops and burns often resulted.

The dream, one she'd recently relayed to her shrink, had been a favorite but it had been causing problems when it turned on her. That one also took place at a deep woods sinkhole where she and a boyfriend, also skipping school, had pried limestone chunks out of the bare wall of the sinkhole and used them to deep dive. The added weight could quickly pull them past twenty-five or thirty feet and into ear-popping pain territory. The mini-boulders would pull them through a thermocline; a horizontal layer that separated cool water mixed with muddy sediment and tinged green from algae to clear-cold black water below. The shock to the skin was more than sensory. The soul-sucking cold reminded her that there were places on earth that humans should leave sacred. Dropping the rocks in that cold, dark realm, she and her friend would face off, cheeks puffed, foolishly daring the other to stay. Usually only a twitch from one or the other would send them both pulling for the surface in a cloud of their own bubbles. The lovemaking afterwards was out in the open on a blanket in the pine woods; exhilarating, life affirming. She'd realized later these experiences were why she had signed up for a mixed curriculum in biology and ecology. She truly loved those pine woods, its critters, and its ever-changing shades of green and russet brown. This morning and too many others, that dive into the cold black of a Florida sink hole had morphed into her reconstruction of the drowning death of her first husband. She'd had to get up and pace to slow her heart. She'd gotten back to sleep again, finally to dreamless dark nothing.

Now, with Carl quietly breathing beside her, she wondered why, with Master's degree in field biology, had she ever been convinced to go into law school and how, after graduating with high honors, had she been convinced to go into corporate law? God, and then family law. *Dammit! I've spent half of a professional career on meaningless bullshit.* She wondered if the money had been worth letting those scumbags hijack her career.

The border of light around the curtains had gone from a yellow glow to pale blue. Dawn had overpowered the security lights outside. She heard a slight hesitation in the slow breathing beside her and called out softly, "Carl?"

"Umh, yeah, what?"

"Are you asleep?"

"If I say no, it's obvious. If I say yes, you hit me for lying."

"Well?"

He muffled a mock groan into his pillow and reluctantly said, "Yes."

In a single move she yanked a spare bed pillow into an arc and thumped him across head and shoulders.

"See?" He rolled away from her, grabbing at his own pillow. She grabbed it also, to prevent its weaponization, and the tugging grew playful. At the foot of the bed, Jazz leapt out of danger and darted for the hall door. In a few moves, she was straddling him, holding a pillow over his head. His arms flopped out to each side, motionless. She peeked under the pillow's edge at his foolishly grinning face.

One eye popped open. "OK, I'm half awake."

She leaned over to kiss him on the mouth, a closed mouth morning kiss. When his eyes opened, they shifted down to her breasts. With her leaning forward, gravity was very kind to their profile. "Enjoying the view?"

She gave him another peck on the lips and rolled off. Staring at the ceiling, she said. "I could get used to you being here in the mornings."

"It's negotiable, as long as you let a person wake up a little more gradually."

"That could work." She edged away from him and pulled a single sheet up over her chest against the morning chill.

"There goes the view. Nice room, no view."

"Careful, I've got more pillows." She sighed; the memory of the dream slipped away but not its chain of thoughts of wayward careers. "Carl, how did you get into what you do? Nice kid takes computer science in junior high, expert Pac-Man pro becomes geek?"

"Not even close." He breathed in deeply.

Is he going to give me the real story or cocktail party version?" She propped up on an elbow, facing him. "Promise you'll never lie to me, Carl. I mean that." Her hand drifted over and found his and grasped it.

"OK, well, maybe at Christmas and birthdays." He allowed, "there have to be some secrets."

"Deal. So?"

"OK, I gotta pee, and we both need a shower. You go first, I'll start coffee." He rolled out of bed before the second pillow plopped onto the floor, missing him.

~ ~ ~

She came into the kitchen to find a bagel warm in the toaster, eggs frying to an over easy finish, and steam rising from her favorite mug. He pointed across the living room to a painting that filled a narrow wall and extended from a few inches off the floor to near the height of the twelve-foot ceiling. It's near-life realism had made him walk close to verify that it wasn't a photograph. The rendition of a kelp forest with the viewpoint approaching vertical at the top and horizontal at standing height. The effect put the viewer inside the kelp at about thirty feet. "That's an incredible painting."

"Thank you. David, my first husband, had it painted from one of his photographs." She sighed, looking at it anew. "I see it every day." *No wonder that dream has been corrupted.* "David was at Woods Hole; those dives were his favorite diversion. David Cecile Bridgeton. He was a marine biologist, PhD. Did what he did for the love of it. I loved him for it."

"There's a lot of past tense in that little summary. You still have the painting. Do you mind if I ask?"

"No, we really haven't swapped life stories. Seems like you've just wormed yourself into a hole in my soul, not that it doesn't feel good. I like you there, or here." She smashed the yolk of an egg and dipped a corner of her bagel into the yellow goo. As she finished chewing the delicious bite, she continued. "He died; an accident. They were down below eighty or ninety feet on hose rigs, and something went wrong on the dive boat. A grad student set the wrong mixture on the compressor. They were too deep and had been down too long for a rapid ascent. They heard it all happen over the mikes. David's dive partner had gotten himself entangled in some of the gear that snagged his hose. He went a little crazy from oxygen starvation and came for David's breather … there was a struggle … both of them drowned."

"God, I'm sorry."

"Panic is a diver's worst enemy. I haven't been in anything deeper than the pool since." She thought of the evening snorkel swims she and David would take, beyond the surf crashing outside on the rocks below. She hadn't even been down to the rocky beach in over a year. Carl stood, apparently lost in the kelp.

He turned to face her; his face painted in questions. "You said first husband, there was another?"

"Yes, a fling turned temporarily serious. It was about a year after David died. I was a little randy, and I made the mistake of confusing satisfaction with love. It was stupid, a sixteen-year-old's mistake. I was thirty-six." She blew on her cup out of habit. It wasn't hot anymore, but the pause allowed her to deflect. "I see it all the time, rebound relationships that don't have any substance." She looked at him pointedly. "It's been long enough since that one went down. I finally have things pretty well sorted out."

They were silent for a while, but as she finished the bagel, she asked. "Any serious encounters of the heart for you?"

"Married, then divorced, one kid–estranged."

126

"Welcome to modern life. Boy or girl?"

"Girl; whip smart and pretty. She'll be nine, no wait. Ten in May. Damn, I should know."

"Yes, you should." She narrowed her brow in concern. "That's pretty estranged."

"Margo, my wife, ex, took her back to France five years ago. She's remarried and cast me as a devil." He picked at a barely existent hangnail. "It's hard to overcome that negative vibe from so far away."

"I don't guess I should ask why…why doesn't matter much in the aftermath. In my family practice I've seen every possible convolution of families on the rocks."

Carl continued his history. "I worked too much, stayed away on business, and…I had one little side trip. I got drunk, high, and stupid all at once. Somehow, the party ended up on YouTube and a girlfriend sent it to Margo. Game over." He studied his fingers, no hangnails to pick at.

After a too long silence, she said, "Life sucks, then you die."

He had been randomly pacing in the undefined space between the open concept kitchen and the steps down into the living room. "There have been a few encounters since Margo, but nothing exceptional till you." She walked over to him and wrapped her arms around his frame.

He pulled back and took her face in his hands. "But life doesn't have to continue to suck, and no one's going to die anytime soon."

"Right."

She turned to the counter, grabbed both coffee cups, and with a nod invited him to step down into the living room. She elbow-flicked a wall switch that opened shades on a brilliant sunlit ocean scene. When they had settled into the couch, she prompted him to go on. "So, where did that story begin, before Margo and the kid? You were going to tell me before we got up."

"Well, you were beating me with pillows." He took a slurp of still too hot coffee and tried to become a little more serious. "This part always feels like Steve Martin's on-screen bio-schtick. You know, the 'I started out as a poor black child' bit."

She waited, knowing he expected a laugh.

He continued in a more serious tone. "I grew up in the barrio, East Los Angeles. Mom used to come over from Tijuana on day work permits as a housekeeper slash nanny. One day, knowing she was pregnant, she decided to stay here, fake visa. Seven years later, I'm in public school, life is good. I went to paradise every day, and they had free lunch. At home almost every night it's either mac and cheese or PB and J." She heard a lump in his throat fighting for breath. When it cleared, he said in a tone that sought humor, "That's all I could cook for a long time."

Her hand squeezed his gently. She had picked up on his authentic Spanish pronunciation of Tijuana with the breathy h sound at the j instead of the Anglo pronunciation with a w. Her brow wrinkled, thinking about her wheat-bread, too-liberal-for-the-local-church-ladies upbringing. "God, I'm not sure I can imagine."

"Some people had it worse. I knew a kid whose mom got picked up in a sweep and got sent back. He didn't see her for a year." His timbre changed, deepening again as he pushed the memories of painful separation away. "But where was I? OK, school. I loved school. Loved science, math, got the computer track in middle school, worked after school in the computer lab repairing and replacing, reloading software. So yeah, middle school geek but with an East LA accent. But later, I scrubbed most tracks of my childhood. I worked really hard to get rid of my wetback accent." He changed timbre and musicality entirely to mock Monterey speak. "Like, you wudda thought I was from Venice Beach. I, you know, bleached my hair even whiter than the blond I started with. Once, like, I even priced blue contact lenses."

"I was wondering; where'd you get that blond hair? No black roots."

"Remember Mom was a housekeeper? Rape is an occupational hazard in that line of work. When she began to show, they kept her on and moved her into a spare room, to make sure she could still care for their kids. When I came out blond, there was major family heartburn! She was put out on the street. Couldn't tell you much about that; I was only a few weeks old."

She took in the disaffected neutrality on his face and decided it was forced, that he didn't want to spend more conversation on the sucking wound of his childhood. *That's why there are mental health counselors.* "Shitty way to treat people. So, where's your mom?"

"She passed, uhm, seven years ago now. Breast cancer, in a county clinic she volunteered at. No insurance." His voice had thickened as unwelcome memories came back.

For half a beat, they shared a look of mutual understanding and beating unbearable loss.

"OK, back to your rise to hackerhood."

"In the nineties, I helped out after school in the school's IT lab. We had antique green-screen DOS machines and then early versions of Windows that needed weekly attention. When I got out of high school, I went to a Microsoft tech school on scholarship. I did so well that I got another scholarship for advanced systems, and the Microsoft school almost hired me. That would have been the end of it. I'd have become a corporate wonk. End of creativity."

"What happened?"

"I had been cruising the gray edge of the dark net and got hooked. Some gamer friends I met in chat taught me how to take my skills further, which sites had keys, hacks, lessons, practice URLs to screw with that didn't cause too much damage. We sometimes intentionally hit our friends' clients if we got short of income. We could bill four hours for a ten-minute fix. I ended up working with two of them, setting up our corporate security company. I was almost twenty. In my own business. No one cared too much what our certifications were, what our age was, what our background was, or even if I had a green card, or needed one. My partner Ollie didn't even carry a forged one, and he was the real deal illegal. Came across for a few months at a time. No one cared as long as we could keep their sysops employed. Our customers only cared about our ability to lock a system down and how to scrub it clean when some jerk opened up the wrong porn page and infected a whole company."

"Guys and their porn."

"Yeah, well." He let the silence admit to his own prior interactions.

"So, then you become Mr. SUPER Hacker." Here air quotes emphasized it as a superhero title.

A grin flashed in response. "Well, yeah. I got pretty damn good. My handle is known pretty much all over the world."

"Handle? You've got a—, what's your hacker name?"

"Beyond ICanD and Gr8efX?" He pursed his lips and drew a zipper across them. "Ahhh... can some things be secret for a while?" She shrugged OK, but her eyes were asking for more. He offered, "It's like asking Mossad if they were trained at Langley, or asking a woman if she's on her period, or if—"

"Got it!" Holding her mug under her nose to breathe in the coffee's last encouraging vapors. She took him in, his loose posture; seeming ready to flee or simply comfortable in his skin. It could go either way. His easy smile could have been trained in the barrio as a get-along street skill or a reflection of his comfort level with her. "Something's missing." His eyebrows raised. "You are too fit for a Dungeons and Dragons, caffeine-swilling, computer cave geek."

"Aikido."

Her blank look stared back, uncomprehending.

"Think Kungfu, almost a black belt."

"Do you still work out? Or is it train? Do you still train?"

"No." He shook his head, expressionless. "Dojo masters traded out about a year ago. The new one was less into defense and more into how to kill your attacker. I took it for personal self-defense. Too much intimidation as a kid and too much pointless death on the streets. But with the new master's focus on killing blows, I was afraid if I was attacked, I'd slip into fight mode and pull my attacker's heart out of his chest."

She whistled through her teeth. Not knowing how to respond to that last bit, she said, "That's quite a story."

"It's not something I dwell on. There were some bumps in the road I'd rather pave over." He lifted his coffee cup sucking, at the last drops. He got up, grabbed both mugs and filling them, prompted her. "What about you?"

"You can guess most of it. Small town girl raised in the church, but rebellious almost to the point of excommunication. Type A personality, 4.5GPA, cheer squad. But the scholarship to FSU was my ticket out. Otherwise, I'd have ended up at the hardware store or the county clerk's office. Could have been a good life I suppose, but I was looking for a way out before I was twelve and knew grades were the ticket. So, BA in Biology, emphasis on marine systems. That's how I met David, in dive school. We moved out here when he got hired at Woods Hole. I took my JD at UCLA, interned through law school at World Promise Fund. I was going to be an environmental lawyer. But I wasted nearly twenty years on litigious corporate assholes. See, not nearly as interesting."

"Bet there's a lot more to know if we were drinking scotch instead of coffee."

She took a long pull on the cup, considering. "Jeezus, lots of happy years for a while. Then got pregnant, redesigned a room for the kid, lost her and some of my girlie parts. No chance of another try after that." Her voice nearly broke, but she continued. "Lots of pain, an affair or two. Well, he had two, I had one. The American story with counseling, reconciliation, and an almost contented slide into my late thirties. We were finally happy with each other. When he died, I thought I would too. Later, on a bender with my gal pals, I met and then way too soon married he-who-shall-remain-nameless and well, shit, Carl! That next marriage was just stupid. Let's leave it at that."

She put down her mug. "Let's go back up the mountain."

"I still need MY shower." He raised an eyebrow in invitation.

She slipped into her Southern girl accent. "Well, just git up an git, then. And don't tempt me with your nekkid body, ya hear?"

14

Chu sat alone in the time gate's workroom. He wasn't getting much done. He kept finding himself stalled in mid-action, stop motion version of life in a vacuum. Grieg was in Neenach on a data run, and Noel? Noel was gone. *Gone! I murdered Noel! How could that have happened?* The horrific reality of the choices he'd made and the betrayal in Noel's eyes that he so often replayed from those last moments had been making him gut-cramping physically ill. Chu pulled some of the data off the last cube Noel extracted from the Chick-a-Saw server bank. This would be Noel's last contribution to humanity. Stop thinking about Noel—think about whether Carl can create the data interface he wanted to develop.

Chu opened clenched eyes and shrugged a shoulder to blot the moisture tracks on his cheek. He'd been upset for a few days but finally thought he had his emotional roller coaster back in check. Grieg had taken the news of Noel Antonides's accidental death relatively well. Chu's explanation was believable considering the slippery footing on the ladder after a rain. A simple fall, a concussion, and then exposure from a dislodged face mask.

Noise at the decon door told him that Grieg was returning. Taking a deep breath and swiping his eyes again to ensure that he was dry-eyed, he waited until Grieg was fully in and had the breather mask off. "How's da weather?"

"Sunny, dusty, hot, horrible. Same as usual." He brushed at his bio-suit as if to brush off dust missed in the decon chamber, and then set his breather mask's filter into its cleaner port. He joined Chu at the work table. "You going over Noel's last data grab?"

Chu mumbled an incoherent grunt that could have been, "Cha, manno."

"He get anything good?"

Chu turned and scanned Grieg's eyes for anything other than a simple question. "Cha, Noel was getting some good stuff from de General Carbon site. Need to put dat on your list for anudda hit." He motioned for a file to open that displayed on a blotter-green holographic background. A buff colored packet opened, fanning out an array of data sheets. "There's also this."

Grieg leaned in, taking in the patent office application. He exclaimed, "Bugger all, that's—"

"—Herbert precipitators." Chu finished. The 20th century writer Frank Herbert's atmospheric moisture collectors from the *Dune* series required a coating that both attracted atmospheric water vapor and repelled liquid water. The fine web work paddles arrayed above sumps collected water in the cool pre-dawn hours for long-term storage and eventual terraforming. The 21st century inventor had named them after the writer. The time station's dew precipitators still worked, despite sand pitting which reduced some of their effectiveness.

Chu found his cheeks dimpling into a grin. It was a welcome change from the grief he'd allowed to overwhelm him only a moment before. "Good find, cha?"

"G'donya Chu! That's terrif mate!" Relearning the coating formulation and application science would be good for the crew back in Wellington. "Let's get that on the next packet out to Wellington, kay?"

"Cha." Chu felt almost calm now. He'd found something useful, or rather Noel had found something really useful for the crew back in Wellington. He asked, "Still daylight?"

"Sun's going down now. Had to get back—battery got too close to red lining on the rover." He looked up briefly from unlacing his outer wear boots. "Sandstorm silted up the roof collector."

"Hm." Chu pursued his ruse, "About that ladder, any chance you could keep an eye out for someting to put actual steps on dat ladder, instead of rungs?" He looked carefully for any tell, any sign that Grieg might think anything but the lie that Noel had fallen from the weather-worn ladder.

"No, I still don't understand quite why he'd go out in a sandstorm. If I'd needed help, I'd have called in."

133

"Cha, he worries too…worried too much." Chu tried to maintain his focus on his screen. He didn't want to make eye contact.

He felt Grieg's hand rest gently on his shoulder, then a light squeeze. "Sorry, mate." Another light squeeze. "Sorry."

Chu appreciated the probability that his lame fabrication had holes in it. He pressed it on. "In de morning, le's go put some more rocks on 'is grave. Keep off de growlers, cha? Maybe tidy it up?"

"Sure, manno."

Chu leaned forward, cupping his eyes in his palms. "And his Rads weren't good either."

It hadn't been too much of a stretch to convince Grieg that Noel had gone out to help Grieg get back in. Grieg's rad meter levels had also been going up, and the two might soon need to rebuild CalStation's overworked blood filters.

After a few moments of uncomfortable silence, Grieg found some filler that was both deflecting and useful. "Well, mate, it's you and me now. I'm going to work on the plasma filters. See if wiping the screens again will help." He turned away and began to dismantle the housing on the plasma inhalator. As he turned the corner, "Tho, prolly gonna have to steam clean the nanoparticulate layer too."

Chu tried not to think about his former lover lying beneath a pile of stones at the base of the retaining wall. It was certainly convenient that there was a ready supply of stones at the back of the farmhouse lot. Lids closed, he could still see the shock on Noel's face as he fell backward. He tried again to put the image away, to not feel the guilt. He'd heard somewhere that almost half of all homicides were by people who knew the victim, and that many of those were partners or lovers.

Noel Antonides would join the ranks of the brave explorers and adventurers who sought the technological answers needed for the remnants of humanity to carry on. Chu had some of those answers before him in the little cube and was determined to pass them on to the past as well as to his own time's research developers. He remembered Carl's mumbling about language interface frequencies. Something about compiler and machine language and octal text translators. Although he wasn't a data geek, he could search a menu as well as any, but Carl's skill set was a bonus. Carl said he had a plan to link two data technologies that were at least 150 years apart. The kid's toy demonstration had been the breakthrough.

Breaking the silence, Grieg asked, "Have you told Wellington yet?"

The question slapped Chu back into the present. "No. No, I haven't. I didn't want to break de news just yet dat a new hero would have to be added to de memorial wall."

"Well, you need to report that we have some good info on the layered battery manufacturing problem. And the precipitators, bleedin' heroic news there. You think you've extracted what they need, right?"

Chu measured the question from a production point of view. "Cha, I tink we have enough now for 'em to be able to set up at pilot scale, but mineral sourcing for de doping metals still needs research. Most de salt flats available in the Western States Alliance and de Manchurian Republic are still too rad. And de Congo, Shyte! Take some brave muthas to go bush bash in dem woods. But de here and now?" He let the aside drop. "Look, it may be useful to have you look for records of mineral deposits in the Chilean deserts. I tink I might need your help on dat trip. Most of dose recen records are goin' to be in Russian."

"All right," Grieg nodded. The plasma filter lay open on the table before him, a nameless goo congealed on the filter bed. Grieg passed a rad meter over the mess and nodded as the needle jumped halfway across its scale.

Chu thought about the differences between him and Grieg and the unexpected visitors, Parker and Carlos. When he thought about it, Chu figured that more than half of his generation was in a bad sack, what the 21sters called damaged goods. The 21sters had no experience of the end of the world. Chu's generation had been brought up in the shadow of a civilization that had seen humanity's zenith. As they matured and understood how dismal things were, how much was missing, many could see total collapse looming a few generations ahead. That was why many of them signed up for the Time Service. It was also why many more of them committed suicide.

Even a short tour of duty in a time gate in a toxic zone could lead to fatal cancers. The hundreds of cures for different cancer varieties discovered in the middle to late 21st couldn't fix the problems that too many splitting atoms in the neighborhood caused. As the mission's comm officer, he'd let Grieg and Noel do most of the heavy lifting down at the data mines. His artificial liver was performing perfectly, but his leucocytes still had a bad habit of attacking healthy tissues. *Nothing for it; I'll have to kick in now, down at the mines.* He twirled a data cube on the table and watched as the physics of the thing made it bounce up onto one corner and spin.

As it rattled to a stop, he gave up on the information dump problems and decanted a tumbler of water. The 21st ers were due back soon, and he'd have to get the information onto the small handful of thumb drives Carl had left. He decided not to worry about it too much. If Carl was as good as he claimed, when they came back, the two of them could figure something out.

Hours later, he was still in a deep funk, trying to consider options on which of their recovered techs Parker and Carl could best use, when the gate's time portal membrane brightened to blue white. They were here.

~ ~ ~

Parker went through first and stepped aside. Carl followed seconds afterward. Both were dripping and wore apologetic looks of concern as they tried in vain to contain the droplets. Chu anticipated their concern. "Don' worre you none about the water. We could use de humidity in here." He decided to tell them about Noel right away; most of it, anyway. "I have bad news."

"Noel Antonides, is gone. He's dead."

"Oh my God!" Parker gasped.

Carl's eyes narrowed in worry. "How? What happened to him?"

Chu nodded to the other doorway. "He went out in bad wedder. He went to see if Grieg would need help getting back from a dust out. He fell off de ladder. Mebbe gust of wind. No one knows. No one to know. Grieg found him ona groun when he come back."

Chu noticed Parker looking at the back door, the airlock door to the decon chamber. She said, "I'm not sure I want to go out there just yet, but I imagine the old farmhouse isn't here, right?"

Grieg answered, "Only a line of bricks where the foundation was. It helped the original Time Service crew know how to place this unit. From the files, it was pretty beat up when the site was set up."

Parker tried to imagine a scene from the last day. "Do you think anything was here — after?"

Chu said, "Whatever was here was blown into carbon dust and spread out across the valley." He nodded to the back door. "And, it's not healthy out there now, pro'bly won't be for several hundred years." He added, "Dere's places out dere will be dangerous for a tousand years, if anyting is still alive to walk whoop whoop out dere."

"He means go on an excursion, a hike," Grieg offered, then turned to adjust gear in the adjoining room.

Parker looked hard at Chu. "You look really rattled, about Noel." Then, with more sensitivity, "Were you close friends?"

Carl bumped her with an elbow. "Park, there's a pretty tight bond that develops among crew members on a mission."

Chu turned his head away from them. He said after a long thoughtful exhale, "Noel and I were more than mates. We were partners. We school chums in de same small town on NZ's nor' island. We were … close."

From the adjoining room Grieg's voice offered, "They were bed bugs, a little friendlier than mates."

Chu's eyes flashed in anger as he looked toward the second room but quickly damped down.

"I'm so sorry for you," Parker said. Chu looked at her, noted her right arm ready to extend comfort, then with apparent indecision, falling back.

~ ~ ~

The visit calmed after a round of Mexican beers and chips with queso were put away. Grieg tossed a memory cube at Carl. "It's at about half capacity now. Noel restored that one from the junk pile. He was clever."

"What is it you want me to find?"

"Your society has a nascent photovoltaic industry. It has efficiency problems that you'll be solving in a few decades with better substrate, but even with the price drop that comes with a 400 percent increase in output per area, you still can't store the power, cheaply." He pointedly looked down at the cube. "In there are the manufacturing and schematic designs for layered carbon batteries. It's what powered the future before the war. Cars, phones, satellites, you name it. Banks of layered carbon batteries the size of trucks were installed for every three hectares of PV array."

Grieg nodded at the little cube. Those new batteries allowed for the power peaks of evening and early morning when the panels were putting out nothing. There's one in this cube that's good for eight hours after only a half-hour's charge in full sun. It will maintain itself and perform necessary data transfers in room light. It will shut itself off to prevent data loss if power levels get too low. It was state of the art sixty years ago. It's one of a few hundred in the world that still work."

Parker's voiced astonished wonder. "You're giving that to us?"

"You need it. Your world needs it. My world needs you to have it." Grieg added, "There's more. You ever read Frank Herbert?"

Carl's face said yes; Parker's looked confused. Grieg continued, "One of your 20th Century writers. He wrote *Dune*? He predicted dew precipitators that could suck water out of the morning air and collect it underground. We found the tech for those precipitators. It could be terrific for areas where you're going to want to stop desertification. Northern China, the Sub Sahara, the central Amazon waste lands."

Carl examined the fine silver fractal patterns in the PV coating and mused, "Amazon waste lands, huh." He handed it to Parker. "How do we drop all that tech on the world?"

"I don't know. But the technology for the battery needs to be put in the right hands. From what I learned in studying your time, I'm recommending Musk or maybe Tyson." Grieg paused briefly. "Or, if you know someone credible with a history of making a public gift of patents, good on ya. It just needs to become public sooner than later so that real solar will get kick-started sooner."

"What about the data cube itself?" Carl asked.

"It's a nice toy. Don't worry about it, and don't share it just yet. The major corporations are already ahead of Moore's law and your civ will keep up." He considered further. "Need to think about who to get that photovoltaic tech to at some time in the future, but not important now. It's going to make the first manufacturer rich before the Chinese steal it."

"Whose law?" Parker asked, a little lost in the thread of the conversation.

"Gordon Moore." Carl said. "He did early work for Intel when transistors were the big new thing. They were just beginning to be miniaturized onto printed circuits. No more wiring. He said that computing capacity would double about every two years." He nodded appreciatively at the little cube. "I guess this is really no big deal for, what? 2150 or something?"

"That's about right. They were everywhere."

Carl squinted and read tiny capital letters in silver at one corner of the base. "25CT, 2,500 terabytes."

"Remember," Grieg warned, "only half of that is functional. It's an old cube"

15

The wall-enclosed, 4,000-foot homestead that Carl had purchased to satisfy goals he'd had as an impoverished youth was largely unused. One interior room, windowless, wrapped in copper mesh and wired to a high-capacity fiber link, was his favorite space. He called it his grotto; equipped with high-fidelity for sound, it was his geek retreat and his working platform.

Carl was puzzled at the central display screen. There were three parallel primary function monitors at eye level, and a fourth security monitor mounted above displaying six exterior views. He straightened his legs, rolling back in his high-back office chair, and studied the evidence developing on the center monitor. With an absent-minded extension of his right hand he dialed down the volume on a high-energy movie soundtrack and focused on the stream of data as it slowly tapped out across the screen. A digit by digit progression of characters halted, a few more were added, alternately as if someone were reading a complex character string and entering them three or four at a time. *What the Hell! Someone is mucking around in my backyard.*

The problem, and it was a big problem, was that someone had access to the Abadon family's private stash in the Caymans and was in there with him. Knowing he couldn't interfere with the other operator's entry without detection, he dialed up screen video capture. He could study the efforts later; there was serious work to do now.

He leaned in closer to the monitor, reading the line of text as it developed. He anticipated that the unseen other would eventually send in a worm and so readied his own worm killer and waited. Carl had never encountered another hacker or presence other than a sysop in these dangerous waters. This was no sysop; he was being hunted!

Here, millions of dollars, euros, yen, rubles, yuan and even dinars were funneled into entirely legal, but usually secreted, accounts. Here the world's wealthiest kept their secret stash; funds from less than legal operations and unreported incomes from legal enterprises. Whenever discretion was imperative, Caymanian bankers and their armed and electronic security experts were among the world's best at maintaining that secrecy.

He should know. He had been trained by one of First Caribbean Bank's best consultants long before he set out his own shingle and years before this current adventure with the lovely Ms. Parrish. What a trip she was! Over a year later and…

"Now!" he shouted to himself. Standing hunched over multiple keyboards arrayed like a rock pianist's stage set, he tapped the enter button with a flamboyant flourish. The other presence had executed the expected hunter-killer program, a worm designed to seek out illegal activities such as the sublime thefts Carl had been committing for months. Carl's new routine invited the worm into an apparent irregularity, waited for it to start reporting, then issued a false report.

It filled the upload data stream with time-killing megabytes of code in Cyrillic, English, and Cantonese, generating a massive encrypted database of meaningless numbers. If it were decoded, it would innocently display transfers from one Cayman account to another, transfers between Russian government and private accounts in Hong Kong and the Crimea. Succeeding in that, Carl's routine would seek out the path and source of the other operator. Hours of preparation for such an encounter had paid off. The long upload time for the database allowed his "gotcha" routine the few seconds it needed to find the other meddler.

"Gotcha!" Still standing, hunched over the console, Carl's fingers danced on two of the four keyboards controlling his systems. He pulled up a reporting graphic displaying a world map. A bright yellow crosshair appeared over the Caribbean islands. Another appeared over Morocco connected by an orange arc. A third jumped to St. Petersburg, a fourth flickered in London and solidified as a final arc connected the now pulsing crosshairs.

"London? I guess that makes sense." The pulsing stopped as they settled to a creamy yellow. He pulled the chair back into position and sat, considering his next steps. "Ah." Jumping up sent the chair rolling back again. Stepping over to a wall-mounted small parts organizer, he withdrew a plastic drawer from an array of twenty-four identical twins. Carl tossed a collection of hand-labeled thumb drives on the small area of his desk not cluttered with drives, connecting cables, empty five-hour energy bottles, and two crumpled, empty bags of chips. He found the one needed and leaned over to push it into a USB slot.

"Well, aren't you clever?" While he'd been searching for his goofball routine, two more crosshairs had appeared. From London, his tracer crossed west to Salt Lake and arced back across the US to New York City, then on to a server bank that appeared to be on top of Los Angeles County. "Holy Mother of God!"

But the last crosshair began to pulse; it wasn't over yet. His planet view shifted eastward and a new crosshair illuminated over the Kentucky-Ohio border. Immediately, another orange arc connected the last two points. "What the—" He waited, to see if the tracer still had work to do. With a nervous jab, he grabbed at a nearly empty can and lifted it high, sucking out the last drops of liquid stimulant. He gave the empty a practiced toss to a trash can. It noisily circled the rim and bounced out to the floor, rolling to a stop beneath his feet.

He didn't notice. The last crosshair just south of the Ohio River wasn't pulsing. "Is that where you're hiding? I can find you!" At last, all of the orange arcs turned green. While he had been waiting for the tracer to finish, he'd keyed up his goofball. He placed a cursor over the last target, slid the control ball forward, and zoomed into a suburb of Lexington.

He almost hit enter to execute the routine when he paused. He got to his feet again to physically detach the ball camera he used to Skype with Joanie at the office or Parker at her house. He muttered, "OK, I'm dark. Lemme see if I can light you up."

In a small window in the corner of his screen, a man's face appeared. Carl took another screenshot to capture the image. Lit primarily by the screen in front of him, the front-lighting provided subtle varying neon highlights to an intently staring operator. A middle-aged man lost in his work. Carl noted the off-round glasses; he was intently sighting through the bottom third of the lenses. Progressives! The man's butch haircut with a hint of flat top suggested military training or latent tendencies. With little background illumination, Carl could make no estimates of size, weight, or physical condition, but the guy's face did not appear to belong to an overweight couch potato.

As he watched, the man shifted screen left and pulled a can of light beer to his lips. As the man raised the can, his eyes widened, then stared in apparent disbelief. Carl took another screen shot of the man's face frozen in shock. Immediately the view from the other computer lurched forward and blanked out. "Laptop camera light was on, wasn't it?"

Carl let out a lungful of air; he had been holding his breath almost the entire time the man's face had been visible. He chuckled. "Busted!" His pits were sweated. He felt as if he'd just been sprinting. Collecting his thoughts, he made a few entries and was gratified that the goofball's IP address had been captured. He leaned over the keyboard again, fingers flying through a long-practiced routine. His penultimate entry brought up MPeck49@ bluegrasscable.net. Carl couldn't help thinking, "What kind of idiot do I have here? First initial, last name, *and* birth year?"

The next tap told him. Martha Peck, 1228 Cherbourg Rd, Lexington, KY. He laughed, a rapid breathless laugh, "That goofball is no Martha." A fast trip to White Pages and he had Martha's son, Oliver. He pulled a now-expanded image off his printer tray as soon as it came loose and stared at the surprised face. "Well, Mr. Oliver Peck, if that's who you are, let's see how good you are. I can tell you this, O-P, you do have some skills, but you're out of your league. And what the hell are you doing on your Mom's laptop?"

When he left his home office a half hour later, he had a public profile built for Oliver Dwight Peck, son of Martha and James Peck; forties, single, divorced, and recently extracted from a prolonged affair judging from the length of the court record and two kids on child support. No Facebook, Twitter, or other known social media accounts. His run-ins with the Fayette County and Lexington city law had resulted in three moving violations: one for speeding in excess of 115 miles per hour and two parking tickets, all paid. Telephone ground line service terminated back in 2015. No known record of cell phone under that name. His deep search through the Veterans Administration's secured records revealed a rapid rise through the non-commissioned ranks as a Navy Covert Communications Specialist and washing out of SEAL boot camp before discharge for disorderly conduct. Prior to volunteering for

SEAL school, he had been assigned to the Navy's Advanced Technical Information Support Systems. That, Carl learned, was a big database on ship configuration and operating systems. On discharge he had obtained IT certifications in basic (ISC2) and Advanced Web Attacks and Exploration as well as something Carl had found humorous — Advanced Ethical Hacking.

In his early training, Carl had also taken most of the same courses.

~ ~ ~

Carl found Parker staring out across a broad expanse of smog-smeared Los Angeles County when he entered her office. She turned at the sound, nodded, and then faced out, westward. "You know, Carl, one of the reasons I picked this office was a slim hope that on a clear day, I would be able to see the ocean."

"You should have picked the south side office. I get fireworks at Disneyland."

She turned back to him and blew a dry Bronx cheer through pouting lips. "Yeah, well it ought to be visible." Although her tone carried a hint of disappointed little girl, there was genuine adult disappointment there too.

He walked up behind her, enclosing her waist in his arms, and leaned to nestle his chin on her shoulder. "My dear, turn just a little bit south of west." He physically rotated her to the southwest. "There."

"What?" The horizon was a faint line of lighter gray above darker gray brown.

He pointed. "See that dark bump on the horizon?"

"Yes." She acknowledged the smudge. "Isn't that an overpass or something?"

"Mah, mah, Miz Parker. You have an office view of Catalina Island. Of course, it's very few residents insist on the more formal appellation of Santa Catalina Island." His attempts to mimic the lilt of the true southerners usually sounded ridiculous, but he tried anyway.

"Jeez, if I ever meet any Santa Catalinians, I'll be sure to make the distinction. Good to know I guess." She pulled out of his embrace and stepped to her desk. "Can I tell you something?"

Carl slipped around the front of her desk and flopped casually into one of the three plush chairs facing Parker, her desk, and the airborne grime outside to the west. "Sure. Shoot."

She smiled at his slouch, ever boyish. Early on she had misgivings about their romance, the nine-year separation between them. At their current ages, fortyish and thirtyish, it didn't matter, and she certainly enjoyed his athleticism in bed. But later, when she was pushing sixty, he'd be approaching fifty and still virile as all hell. She blew the thought away with a puff of exasperation.

"When I was home, we'd travel to Tallahassee to shop, hit the malls, take in an afternoon movie. Sometimes we'd go play tourists in the big city. They had a great natural history museum there with displays for the kids. Uphill and a hell of a lot of steps was the brand-new capitol building. There's an observation deck up there on the twenty-first or second floor that visitors can take straight from the lobby. The views were spectacular. Most of the city is hidden in its own trees, but the hills roll out in four directions, and there were little placards that gave simple descriptions of what you were looking at. Unless it was a really hazy summer day, and that haze was humidity not dirt, you could see the Gulf of Mexico to the south. The freaking Gulf of Mexico! That's twenty-four miles away, and it was almost always visible."

Carl absorbed the mini-lecture with a sympathetic ear but didn't respond immediately. "So, it's polluted here. Everyone knows that."

"Everyone accepts that." She paused as a medic chopper passed nearby, too low and too close.

"Been that way since I was a kid," he answered. "Summer especially. The heat holds in the exhaust of ten million cars, and the air turns brown. No mystery."

"You're saying that if Tallahassee had ten million cars, it would have been the same?"

He chortled at the thought. "If Tallahassee had ten million people, it would stretch from Georgia to the coast and have twelve-lane freeways like we do here. And dear, that humidity would hold even more of the dirt and exhaust in the air."

Parker thought back to that image on the CalStation viewer that showed Tallahassee's urban expanse fully developed to the Georgia line, but truncated to south by the new edge of the Gulf of Mexico. She sighed and sat, sagging into the arms of her high-

145

backed chair. Her mood was dark even though their new building conversions were almost completed. The new HR director was adding new social media talent every day. Contributions to the new environmental start up were rolling in well beyond their early projections. Still —

"I'm pissed!" she said flatly.

"Wow, I might never have guessed."

"Our little adventure we have planned for North Florida ought to be a beginning."

"You still want to embark on a life of crime and environmental terrorism?"

"I want to incorporate it." Her eyes roamed the ceiling tile grid as if mentally scrolled through her initial list of options. "If we don't leave a calling card, everyone will blame ISIS or al-Qaeda, or the next iteration of mis-informed terror." She stood, stepped around her desk, and took his hand. She pulled him to his feet and pulled into him, pressing against his exercise-hardened chest and looked up into his beautiful eyes. He swallowed, enjoying the pressure of her body. One of her thighs was pressing between his.

He thought to at least finish this chain of thought before moving on the invitation. "Well?"

"What about Verts Sans Frontiers?"

He frowned. French wasn't his strong point. "Uhm, Vertical without Frontiers? What's — "

"Verts Sans Frontiers is Greens Without Borders, like Doctors Without Borders. It will be the voice of a revolutionary movement to point out the fallacy of dependence on fossil fuels."

"It does have that international flavor. People would be less likely to look in our corner of the planet for who-done-it." He nodded his head in apparent appreciation. "I like it."

The sparkle in his eyes either said his buy-in was genuine or she was holding him too close. She thought, *This guy is a keeper. For as long as he's available, I'm available.* She couldn't see further than that for the immediate future. Too many moving parts in their lives had begun to spin, and she had a lot of staff to bring up to speed. *Time to take my mind off the too-long list of to-dos.* She turned back to face the southwestern grime she had to accept as the horizon. She said, over her shoulder, "Come here."

Carl stepped beside her.

"No, when you came in, I think you were behind me."

In obedience, he sidestepped to stand behind her. He put his hands around her waist and pulled gently, pressing her rump into his groin. She closed her eyes, shutting out the disappointing view and pulled his hands up to cup her breasts. "Much better."

~ ~ ~

When her desk buzzer sounded, they had just finished "breaking in" her new desk. It had done just fine, as did the plush black leather couch, and one of the chairs.

She tried to calm her breathing as she answered. "What is it Meg? I'm a little busy."

"Sorry to interrupt," Meg's mechanical voice returned. "I have your tickets to Mobile ready, car reservation is set for pickup at the airport, and your reservations at the Marriott."

"Thank you, Meg." She was reaching behind to clasp her bra.

Meg's voice answered. "It's getting late, Park. I'll leave them on my desk, next to my inbox. OK?"

"Thanks, that will be fine. Have a good evening." She opened her arms to accept Carl's hug.

Grateful to still be young enough for spontaneous sex and old enough to be good at it, she said in his ear. "What was it you wanted to talk to me about when you came in, oh, an hour or so ago?"

"Well, that's a real change of pace," he said, fastening his belt.

In response, she just flashed a "who me?" grin and leaned back against her window.

He continued, "I was at my home office earlier, doing some deep diving into the Abadons' financial transactions."

"Did you get some, uhm, divert some more assets?"

Carl settled into a corner of the recently abused black couch, water bottle in hand. He took a pull on the bottle before starting. "There was someone in there. Someone else was in the Abadons' files."

"Routine transactions maybe?" She was hopeful.

"Probably security. I don't know for sure, but I'm pretty sure. But the guy was going into places that are normally ignored. When account balances run into the billions, odd millions can be overlooked. But if you are looking for lost millions? Well, that's what I'm worried about."

"Jesus! You should stop immediately. Go burn your computer, toss it in a canal."

"Not quite yet. It's easy enough to change an IP address." His smile reflected smug satisfaction. "I rattled him something good, and I have ID on him. His name is Oliver Peck, lives in, or near Lexington, Kentucky."

"Dammit, too good to be true." She waved a hand, encompassing the office, but by implication all of Room for Tomorrow. "None of this would exist without those funds from Abadon's offshore holdings."

"Or PeMex, or Chained Industries, or... don't worry, we're diversified." He held his palms together in prayer. "I nailed his little invasion to the floor. Sent it a false report; it shows all of our deductions as transactions routed to Banc Suisse Cayman. They have multiple banks; it looks like a transfer to one of their formerly secret accounts."

"You're sure your tracks are covered? I'm too young for life in prison."

He paused, taking a deep breath and blowing it out slowly through puckered lips. "My dear, I'm better than he is. He now thinks—"

"He won't know who's been looking?"

"Yes, he will, and when he settles in on the spy, the entity that just did deep records searching on him and his history, he'll find an anonymous account in an ancient stone building, Unit 26165 in Central Moscow. Dead end."

"Right." Thinking about twenty-five to thirty years in prison had a way of taking your mind off the recent sex high. "So, let's talk about our trip to Florida. I want to make sure you meet my mother."

16

The dark green cargo van merged onto the four-lane west of Tallahassee, as unobtrusive as an egg in a hen house. Parker had to agree with the street smarts that Carl had shown on many occasions. In any given situation it seemed, his network, when expanded by the people his people knew, put them in contact with an element she'd probably not want to know. His last contact though had been just as advertised—well-armed and happy to sell to the right buyer.

Riding in the back, securely wrapped up among simple articles of clothing, were a half-dozen claymore mines complete with specially rigged detonators designed by a Desert Storm vet who'd lost a foot to an IED south of Fallujah. Undeterred by the need for a prosthetic, he'd signed up with a contractor doing bomb squad work and done two more years in the sand box. He knew how to make and disassemble almost any configuration of explosive found outside the green zones. All he needed was a cell phone, a few wires, and the cleaning fluid aisle in a lawn and garden center. With real weapons, he was seriously dangerous. His skills were now being carried into a new battle. The code of his calling being what it was, he only needed to know that no one was to be directly impacted by bits of the debris. It was an extra stroke of luck that he had an RPG-29 Vampir grenade launcher and two TBG-29 thermobaric rockets.

Parker had taken Carl to the site earlier in bright daylight and, disguised as a happy couple picking wildflowers on the edge of the road, they found the expected markings. 30" – 24" and two marked 27". These were conveniently marked on the edge of pavement. Twenty feet away near the barbed wire fence protecting the Florida Gas Transmission's pipeline right of way stood the yellow taped markers warning idiots to call before they dug. Only an idiot would dig a hole in what was very obviously a major pressurized gas pipeline right of way. Right? That's what Carl had said during the afternoon visit. They had dinner at a simple but tasty Chinese buffet in Quincy, a small county seat half an hour west of Tallahassee. They'd made small talk while the sun took its time to set. The required post hole digger had been acquired the day before in a rural hardware store miles away in south Georgia. No need to leave obvious tracks. The truck itself was to be expendable and had been purchased cash through Craigslist in Mobile.

They pulled up to the chosen right of way, saw the compressor station in the distance behind its protective fencing, and pulled forward so the dark green box of the van wasn't as visible in the light of the Florida Gas Transmission compound. "Go ahead and pull off up there. You can back into the tree line and be a lot less conspicuous."

Carl did what he was told but asked, "I thought you said no one ever used these roads at night."

"Well, practically no one. There's no reason to announce via a random witness that the local law should BOLO a green Ford cargo van." She'd driven these roads as a teen and knew how tunnel vision in the miles of planted pine can be distracted by almost anything that breaks the pattern. With the well-lit fencing of the compressor station's perimeter just ahead, a dark truck backed into the overgrown brush in the pines would be a lot less noticeable than almost anywhere else in the miles of planted pine plantations.

Carl acquiesced. "All right, you are so right." With her guidance, he backed into the tree line about two car lengths, and they stacked a few fallen branches in front to cut down random reflections from the grill and headlights.

She pulled her red lens flashlight and then pocketed it, realizing that she would not need it between the pale glow of a half moon and the distant lights surrounding the compressor station. She pointed out the two white lines bracketing the position of the 30" pipeline, and Carl took a stab with the post hole digger. He got about a half an inch of hard packed clay for his effort. "No, over here," she hissed and then realized that over the din of the turbine compressors that she could have shouted. She pointed to the marker at the fence line. "There's less likely to be any protective cover here at the fence. There may be a concrete cap near the road."

Leaning on the post hole digger's two handle tips, he asked in some amazement, "And you know this how?"

"I don't—it just makes sense that a phone crew or something might be active closer to the road that 6" from the fence."

"I'll grant you that." He took a 4" deep circular wad of weeds and sandy soil with the first strike. "It's a lot easier digging over here." He went down about two feet pretty quickly and gave the handles to her. "Here, widen the hole so I can extend the handles to go deeper. I'll take it down to the pipeline."

She did, leaving a bit more of the soft sand and chopped vegetation on his pile. "How's this?" She punched the handles down and demonstrated how wide they could get.

"Careful!"

She mimicked him. "Careful, he said to the chick digging a hole on top of the high-pressure natural gas pipeline."

"Right, I want to blow up a pipeline not you."

Carl took the handles, and after a few more tentative strokes, they finally heard a dull reverberating thud from the softened bottom of the hole. They set the first charge in the hole and walked toward the marker for the 27" pipeline.

"DOWN! Now!" he nearly tackled her before she understood the danger. An off-road vehicle was approaching along the inside of the compound's perimeter fencing.

From a few inches off the ground, the damp smell of earthy, weed-covered soil filled her nose. The crescendo roar of the vehicle's struggling engine paused at the nearby corner of the compound and diminished quickly as it had come. She blew a sigh of relief. "What's he doing out here? Do you think they have noise detectors?"

"No, it's closing in on midnight. They're probably doing an end of shift perimeter patrol." Both were still in the damp weeds next to the fence. They rose to see the round taillights of a departing ATV already fifty yards away along the road frontage of the compressor station's fenced compound. Strains of country and western and a slight smell of tobacco lingered as patrol vehicle rolled slowly and unhurriedly away.

Parker got up, dusting her jeans off. "Crap, I almost pissed myself."

"Me too, but I got it worse."

"Worse, you did piss yourself?" She couldn't help laughing. "You're not getting back in that truck with me if you did. You can stay here and direct the sheriff to the scene of the crime."

"No, I set my hand down on a nettle, and it's stinging big time." He shook the afflicted hand as if that would help. "Damn! That hurts."

"Hand me the post hole digger; that's going to need tweezers later." She reached the depth of the 27" line and was about to set the charge in that hole when pine trees around them began to glow. They managed to drop flat again just in time to avoid the headlights of a pickup heading south. It slowed and pulled up to the gate of the compound. A few seconds later, the drop arm gate controller lifted, and the pickup entered the compound. It soon disappeared behind the double-wide that served as a guard shack. "Yup," she whisper-shouted, "shift change. We should expect more."

Standing again, hearts pounding, Carl dropped the second charge into the new hole, and they moved back to their truck. Parker pulled a suitcase out of the way to allow Carl access to the molded plastic case that held the Viper. Releasing the catch and opening it, he blew softly through his teeth. "Would you look at that? That *is* the real damn deal." He opened a small canvas shopping bag and grabbed one of the grenades.

"HEY!" The voice was too close. "Hey, What chall doin' over t'thar?" A very thick Southern male accent was definitely hailing them. The speaker, in a company shirt, jeans, and cowboy boots, was on the other side of the van at the corner of the fence.

Carl poked his head around the side of the van and peered into the stabbing white light of an LED flashlight. He yelled at the light, "Just a minute!" He turned back to a wide-eyed Parker. "Crap, follow my lead," he whispered. He called out over the van. "We're just leaving!" Then loudly enough to be overheard, he told Parker in the worst Southern accent he'd ever heard faked, "Mary Ellen, git your drawers back on and git in the truck. This place ain't as private as you said it was." Then directing his voice back to the compound fence, "Sorry sir, we'll be outta here in a jiffy."

"Hold on now, I'll be right over."

"No problem, sir. We jes' be leavin."

Still unseen, Parker got in the sheltered passenger side while Carl, keeping his head low, got in the driver's side. With an unimaginably loud roar, the ignition kicked the Ford to life, and they burst out of the tree line. Over the noise of the engine and the omnipresent whine of the compressor building, they couldn't hear the laughing of the security guard. They couldn't see much either. Adrenaline had narrowed their focus down to the cone of white headlights ahead of them as they raced north toward the sleepy midnight civilization of Quincy, Florida.

"Slow down. He can't be following us!" She put a calming hand on his right arm. "We're still good here."

Carl lightened pressure on the pedal, letting the truck drop back down to just below the speed limit. "That'll be a story for your mom."

"Carl Reyes! Don't you even think about it."

"Yeah, I don't guess we could tell her what we were doing out here."

"Not in any kinda story, you hear!"

It might have been the pine woods or the location back near her roots, but the Southern cracker girl twang from her North Florida upbringing had taken over her urbane Southern California, non-accent. He stared at her for a few seconds.

"Yes, ma'am. Wouldn't even think on it." His accent was far less genuine.

Parker looked at her watch. "Shift change will be over at midnight. Hour and a half, maybe more, we'll go back, lights out. We'll turn around so you can get a bead on that big turbine building, and we'll light up the neighborhood. I'll have the phone dialed to the last number, and poof—tomorrow there'll be a heck of a lot less natural gas in the Sunshine State." She looked over at Carl. He was illuminated by the dash lights, his attention shifting from the road ahead to trying to pick nettle spikes out of the meat of his thumb. She started to giggle girlishly.

He turned a curious face toward hers. "What?"

"Mary Ellen, git your drawers back on and git in the truck." She broke out into unrestrained adrenaline fired laughter. Giggles mixed with "Git your drawers on" continued for a little while before he joined in. Tension lost to laughter. Then he set his palm back down on the steering wheel and winced in pain.

"First, we're gonna find an all-night drug store and some damn tweezers."

~ ~ ~

At 2:30, they returned and cruised by the compressor station at just under the speed limit, checking for bobbing lights of the perimeter patrol's jeep as they passed. She turned the van around a half mile south of the installation and, lights out, pulled slowly past the guard shack that was set back a hundred yards from the county road.

They pulled over near the fence corner, just north of the several acre compressor station site. "Carl, see if you have a clean line of fire." He carefully set the loaded Vampir on his knee, powered up the rangefinder and sighted to set the distance He then set the weapon to his shoulder and centered the middle of the building with the big exhaust stacks. With the truck in the middle of the road, he had a line of sight over the perimeter fence toward the top of the huge building's sheet metal wall. He could see shimmering heat waves above the stacks even at night. These were big machines. In conjunction with other stations on the line, the compressors maintained enough line pressure to get natural gas from East Texas to Miami. She took her purse from the floor and found the burner phone with the big "1" written in black marker. With a trembling hand tapped the only number in the cell phone's speed dial. "Carl, you ready to do this?"

"Miz Parrish, I'm as ready as I'll ever be."

She tapped the glowing green phone icon. There was no sound, but five miles west on another dark country road, the charge on a well-marked pressure reduction valve station went off. They could not see the yellow-orange ball that blossomed beyond the pines, but a soft orange glow illuminated the humid haze to the west. Several seconds later, a low whump sounded from the distance. They waited.

A full minute later, the guard shack erupted in pandemonium. Red flashers and klaxons sounded, and the few night shift maintenance workers ran from the massive compressor building toward the guard shack. Parker and Carl watched through the pines as two and three at a time the assembled crew leapt into trucks and headed out to the scene of the blast.

A single figure remained standing in the flashing red light of the guard shack's yard. The only employee they'd seen with a tie threw his hard hat to the dirt in a pantomime of frustration, kicked it twenty yards across the parking area, and turned back into the guard shack. Carl, with the RPG launcher readied, leaned on the truck door frame for stability. He took a deep last breath, held it, and slowly increased pressure on the trigger. Bam! Whoosh! The noise and heat blast of the rocket surprised him. He had barely recovered his footing when the RPG's trail ended in the sheet metal wall of the compressor building. The blast was smaller than he'd expected. Maybe it spent itself on structural steel and nothing was going to happen. He recovered his senses and jumped through the open side doors of the van. "GO! GO! GO!"

He didn't have to yell, she had gunned the accelerator, hurling Carl against the rear doors of the van in a heap. They just cleared the pipeline right of way when the building erupted in a brilliant flash that lit the night. As the rising column of hell's fire blossomed into the black, the van disappeared in darkness, heading south.

~ ~ ~

Carl crawled forward, retrieved the other two phones from her purse. He activated their calls. Behind them, in the hole over the thirty-inch line, another phone came to life. A half second later it answered with a claymore blast that split the steel carrier pipe open. A jet of high-pressure gas gushed into the cooling midnight air. It was followed immediately by the second blast from the twenty-seven-inch pipeline. In a blast that could be heard in the county seat, ten miles away, both pipelines ignited in a brilliant white-yellow blossom. The two jets of flame rose in unison, joined into a hellish roiling mushroom in yellow-orange. Fed from the pressurized pipelines, the rising jetting flame began to draw in cool pine land air. The roar could not be heard over the straining engine of the escaping van, and its two occupants could not stay to appreciate the beauty and awe of the fire tornado that was to burn for the next two hours as the pipeline emptied its pressurized load into the rising column of smoke. Pipeline security in the United States, or lack thereof, would never be the same.

The trained for, but unexpected, worst-case scenario had happened. The hatless shift supervisor jabbed in angry overreaction at the buttons on the guard shack's phone. But no signal was going to get out over that line. It had melted seconds ago in the blazing inferno at the edge of the road. Realizing the line was dead, he grabbed at his pocket. His pudgy, fattened palms seemed unable to grip the cell phone from his jeans. By the time he'd convinced someone at 911 dispatch that there was an emergency, neighbors two miles away had already reported an explosion and fire at the pipeline's compressor station.

~ ~ ~

As first responders from three counties began to roll out of their garages, Parker and Carl were, by then, moving south at only a few miles an hour over the speed limit. Their first planned stop, the small golf course motel south of Crawfordville, was to pick up their rental, a blue-gray BMW roadster. The second stop was a high school favorite sinkhole. She knew it to be over eighty feet deep and the resting place of at least three other vehicles. The van was destined for a watery grave in that back woods sinkhole, never to be seen again. To complete the cover for their Florida visit, the two new environmental terrorists ate a light breakfast, played nine holes badly, and headed to the coast for a long overdue visit with Mildred Parrish and her new fiancé. Parker was going to get a new stepdad.

Traffic on I-10 between Tallahassee and Pensacola was usually light at mid-morning. This morning was no different. Their two-day stopover visit allowed for the FHP to be cautious about taking down the roadblocks and conclude that the saboteurs were out of the area. The talk show news of the afternoon was the terrorist sabotage of the pipelines. "Carl?" She turned to him; he was, as ever, nose deep in his tablet.

"Yeah, sweet?"

"That was fun—the pipeline I mean. Not so much meeting Mom's new beau."

"Teddy was alright, I suppose. If Mildred's happy…"

"Sure, but what I'm getting at is the actual act. Digging those holes and blowing the pipelines was fun, no, exhilarating." She thought back to the pounding in her chest; the exercise of digging the post holes accentuated by the fear of discovery; their escape along seldom used country roads. "But."

Carl stopped tapping, looked up. "But what?"

"I'm not happy about the van."

"Well, we couldn't have hung around until dawn to see if we'd covered the tracks. I mean, it's likely that there are scrapes on the limestone at the edge and someone might notice the broken bushes eventually. But we talked about this. Who's going to put together that meager evidence and the pipeline incident? Who goes there, right? You said mostly high schoolers skipping afternoon classes."

"We need to figure some other way of getting rid of cars. Maybe sell them to a junk dealer."

"What are you getting at?" He was trying to connect some dots that didn't seem to have mattered in the planning stages.

"It's the sinkhole itself. I know there are other long dead cars down there. But ours shouldn't be. Crankcase oil, gear oil, dammit. There's going to be an oil slick there for years."

"Hmm, you think that will lead someone to yank it out of there? Maybe use it to trace back to us."

"No. Those woods are national forest. No one really gives a damn except the corporations which farm the land for slash pine. It's just that we polluted a really pretty spot for kids to go enjoy an afternoon and a beer or two."

"So wait—" His face betrayed confusion. "We just blew up two of the four pipelines feeding natural gas to the Florida peninsula and you're worried about polluting a sinkhole?"

Parker looked over at him, saw the quizzical look. "Well, yes." She had to admit, comparing the two acts hand in hand weighed significantly more on the hand holding industrial sabotage. "But it's such a nice swimming hole." She had been thinking about the recent visitation of dreams about her high school days at the area's more popular sinks.

He started laughing first, she soon joined in, she couldn't help it. Despite her frustration, she found herself shaking with a giggle that soon erupted in echoing laughter. She tried putting herself back in her teens, passing a joint or a beer, daring each other to skinny dip, and remembering those long-ago jump-or-dive stolen afternoons. But Carl's humor took her over. He took her right hand off the steering wheel and wove his fingers between hers.

"Sweetheart?" His tone was colored by the hearty laugh attack. "We're going to be OK. I believe it in my heart that we're doing a good thing overall."

"You're sure the press release can't be backtracked?"

"No, I'm sure it can." He had to suppress another laugh. "All the way to a fourth-floor walk-up in the Fourth Arrondissement."

"Paris?" Her smile widened. "That's rich."

"Can't blame the monkey wrench gang for everything, right? No reason the Verts Sans Frontières can't be a little classy, right?"

She lifted their finger entwined hands and kissed his knuckles. "I suppose I'd rather be classy than desperate."

Paris triggered an old memory, plans not fulfilled with her husband. "Carl, promise me we'll find a reason to go to Paris some time."

~ ~ ~

The roadster allowed them a leisurely return to California. With the exception of a few high-speed episodes in the long empty stretches between El Paso and civilization, they averaged only ten miles an hour over the posted limits. When they reached the western terminus of I-10, three days later, both the heat of the fire and the focus of the press had cooled down. For weeks afterwards, commentators, investigators, and pundits debated the identity and origin of the two black-clad saboteurs in a dark colored van. The conspiracy networks came alive with targetless invective against whatever or whoever had staked the small metal sign into the ground on the opposite side of the road from the gas-fired inferno. Who was behind Verts Sans Frontières?

17

Less than a year after the first meeting with Chu and Grieg, the four floors of Room for Tomorrow's headquarters had been furnished to capacity with phone stations and social media outreach stations, mail sorting equipment, grant writers, and project leaders installed in the major research and tech-push areas represented by the twelve board members. The dream had become a reality. But Verts sans Frontières had just hit the headlines.

Television networks declared the last of the Florida Gas Transmission's pipelines had been severed in the quiet backwoods of Liberty County. With thousands of acres of planted pines and a nighttime Sheriff's shift of one, it hadn't been hard to plan and execute the second hit. Carl's network from his old neighborhood had come through. There had even been choices of cartel-trained experts and those with three-tour Afghanistan credentials. In the end, even with "enhanced" surveillance promised by every spokesman for the Sheriff, Governor, FBI, NSA and anyone who wanted a twenty-second shot on Morning Joe the countryside had assumed that something like that could and would never happen again.

Texas and Louisiana joined gas execs in demanding for heads to roll. Cries for lists of Islamist radicals in the Southeast filled the hourly news updates that had little actual breaking news to report. Orders for coal trains could not immediately respond due to their deployment to the shale oil regions in Canada, and pressurized tank trains were serving the fracking boom fields in the Dakotas. Coal would come but not in significant quantity to keep Floridians cool in the warming months. Many of the retired coal-fired burners had been retrofitted to gas, and their complex conveyor systems had been removed. It would be weeks before full power deliveries would resume. Power company execs across the state were apoplectic in their demands for dual fuel boilers. Meanwhile the governor asked that home thermostats be set to eighty when people were at work and for office systems to be turned off at night. The demand for PV solar panels soared to record heights in the immediate aftermath. Parker smiled at this. This was progress; but would it hold?

Room for Tomorrow, the new environmental action start-up on the West Coast just beginning to make waves had been in some of the discussions, but only as a tangential topic. RFT's press releases decried environmental terrorism. Parker's small media relations team published memes, trolled Twitter accounts, and in any way possible deflected from or blasted the actions of the infamous Verts. Any actions it took on the media feeds were mere pond ripples compared to the surging tide of headlines castigating Verts Sans Frontières on the major talk radio and news channels.

Parker knew for RFT's initial groundswell to move to the next step, GoFundMe financing ops would be far from adequate. They might help, but foundations had to be laid. In addition to funding secured through Carl's darknet skills, Room for Tomorrow's public foundations were built on a green email database comprised of the tens of thousands of twenty-five-dollar and up contributors to Green Peace, Sierra Club, the Wilderness Society, and other conservation lands "Friends" groups. Nothing in RFT's organizing efforts suggested that Room for Tomorrow was anything more than a rapidly growing and particularly well-run and well-funded advocacy group with a focus on funding leading edge research.

Carl's data efforts resulted in the hiring of a team of recruiters who found volunteers to create contact lists of liberal-leaning and Green Party supporters to find political activists who might form local fund-raising and advocacy groups. Room for Tomorrow began to make its own headlines as a group that was making phenomenal progress in membership recruitment and financial health.

Besides membership growth and deciding which new projects to fund, Parker and Carl had other problems to think about. The explosion of the Russian pipeline between Kurst and Kiev in the Ukraine was timed to blow while they were airborne to the wind-swept plains of Nebraska. When two unmanned pumping platforms in the Mumbai High oil field were turned into huge Roman candles, they were conspicuously discussing Gulf

Stream electric turbines in Palm Beach, Florida. Verts Sans Frontières metal placards had been conspicuously placed in all locations.

Negotiating the destruction of the Kurst pipeline, and especially framing a third-party scapegoat, had been relatively easy considering the partisan tensions in Ukraine. Russia's President Putin scoffed at the idea of an upstart green terrorism culture and blamed Ukraine's defense forces. Deep irony lay in the fact that Russia bankrolled its military operations in Ukraine and the Baltic with the same oil that passed through its many pipelines. He was less sanguine when sabotage at his oil fields destroyed the pumping stations that supplied many of those pipelines. When the news broke of the May Day destruction of the LukOil tank farm's pump complex in Vysotsk, the world's military bomb throwers had gone to the brink before Verts Sans Frontières had taken credit.

Instead of the cumbersome mouthful *Verts Sans Frontières*, the media had taken up the easy to pronounce VSF acronym. VSF had taken its place on the dark green edge of the environmental movement. Although there was no discoverable connection between the operatives who'd been paid to damage those facilities, no one would ever find fingerprint matches. Genome testing had determined that there were two identifiable nationalities involved; one was ethnic Chinese, the other Russian. FSB and NSA channels opened in a rare show of international cooperation to no avail. The last surveillance video captured of the two suspects showed a tall Asian, and a Caucasian in hoodies passing under a streetlight near the Bahnhofstraße in Wiesbaden, Germany. But the trail died there. DNA material had come to a dead end with no matches. The two suspects would not be born for another 150 years.

~ ~ ~

By fourteen months into the creation of Room for Tomorrow, Parker exercised the option to take the entire stack of upper floors with phone and internet chat stations now occupied five floors of the six-story Anaheim office. The new expansion saw replacement of planned phone bank platforms with Facebook, Tumblr, Twitter, and Hootsuite specialists. Plans were in place for public display and conference space on the bottom floor. The current tenant was provided ample financial incentive to hasten his search for an alternate location.

The board of directors had decided that if funding drives met targets, European and Asian headquarters should be considered. The next funding drive was very successful. Between public crowdsourcing income and private involuntary donations from PeMex, Gazprom Industries, and the Abadon offshore accounts, the capital campaign of mid-summer 2021 was phenomenal. By August, Room for Tomorrow's 200 employees spread between phone banks, chat rooms, email warriors, Facebook farmers, and industrial design teams. Some of the latter had left the robotics design team at Disney engineering to come to work for a better future and a better paycheck.

Parker's favorite projects, by far, were the new Room for Tomorrow rooms being planned for the six Disney theme parks' Tomorrowland venues. Their interactive displays would highlight technologies of the future, renewable energy developments, carbon-electric research, reforestation, and the need to conserve remaining petrochemical oil preserves for future generations who would need them to produce recyclable plastics.

18

The eight-story sandstone and glass edifice of Abadon Industries stood on the Kansas flatlands as a monument to industrial capitalism and Ayn Randian-inspired power. Its façade lent weight, substance, and grandeur to the power behind the Abadon name. With less than thirty percent of its billion-dollar wealth invested in energy production, the family firm still controlled one of the largest sectors of world production. Getting fossil fuel out of the ground and into power company boilers or chemical refineries was serious big business. The company's worldwide reach extended even into the corrupt hallways of the Chinese hegemony. In fact, hegemony as a form of government was a model very attractive to the Abadon political agenda. Achieving this model was a chief motivation for the multimillion-dollar donations to ultra-conservative political races across the country and across the globe. Room for Tomorrow was about to make a frontal attack on American corporate policy. Grieg had drawn short straw and now stood amid the flag poles and fountains at the Topeka headquarters.

Grieg had thought getting access to the building would have been difficult. He couldn't have been more wrong. He found that approaching through channels, as a prospective employee, he was welcomed in with a friendly smile. His artificial vita brevis for Griegor Pavlukin, described a refugee from Kharkiv, Ukraine, as an orphaned teen and mathematics prodigy accepted by virtue and NATO intervention into Germany's educated technical elite.

His forged certificates from the Technische Universität of Berlin backing his ability to analyze complex systems with intelligent algorithms were supported by ersatz class records inserted by Carl into the University's data center. He had been initially unimpressed with the on-site security at Abadon. If all you have to do to get into their guardhouse was to be an employee, then what could be easier? Finding out that his new-hire status gave him almost no data access privileges upped his assessment. His computer, he found, had limited access to corporate servers. Grieg's workstation was on the research server with no cross links to places he'd like to investigate.

The RFT board decided he should go proactive. In his second week, Grieg produced a world-view analysis of remaining crude reserves, emphasizing his conclusion that most consumers were not concerned about a pattern of use that wouldn't experience extreme shortages for two more generations. Considering the political climate on the "inside," he couched the report on the best times to invest in alternative and renewable energy resources and change-of-state energy storage technologies that would make wind, solar, and tidal sources more feasible. The report indicated that research on those fronts should be stepped up due to the undeveloped state of those fields. He noted that long-term portfolios on petroleum assets should be considered to be refocused on advanced plastic hybrid materials. His appendix included the research universities with the most likely programs.

In appreciation of this work, his managers had a heated debate over whether to fire him or give him a raise and increased access. His fourth-week report on key locations to install ocean current turbines that would not be affected by a potential loss of the polar ice cap. The report was titled *Money for Nothing*. It was a hit and resulted in early talks with Japan on replacing all nuclear capacity with ocean current turbines. That report was followed in his sixth week by a report on reuse of very deep oil wells as geothermal energy fields. That raised eyebrows and interest in the boardroom. It was keyed to an interactive GIS map that displayed those locations and the regions that had the best geothermal characteristics, keyed by those most easily accessible which were already under ownership or lease. The gold star locations on this map indicated well fields that were nearing exhaustion. The graphics included adjacency to high capacity electrical grid assets.

The next project, current research into the development of laminar carbon batteries, had the board looking for investment options in the field. It was salted with a breaking theory outline taken from very real research that would have been initiated by 3M in 2035. Room for Tomorrow saw this as an easy way to share information already offered to Tesla Motors, GE Labs, and JPL. When the Abandon section supervisor took Grieg's work further and found these institutions working on the cutting-edge tech, his report gained instant credibility.

His last project, before resigning after only three months, was another white paper on the long-range cost benefits accruing to regional stabilization of satellite feeds in Arabic and Farsi to the Muslim Middle and Far East. The feeds would supply children's programming that promoted literacy, women's rights, and modeled smaller family sizes while countering the culture of religious extremism. He wasn't sure if anyone on the board actually invested for a term longer than next year's annual reports.

In only thirteen weeks, Grieg had accomplished his mission, sowing seeds of solid eco-industrial research, firmly tamped down in the questionably fertile soils inside the Abadon tower. He left on his machine an apparently unfinished but perfectly useable white paper on the need to provide Faraday shielding for significant power grid assets, transportation and financial infrastructure, and all new telecommunications satellites. If leaked, the same reports could be used as public relations fodder in future negotiations with the corporation's leadership.

~ ~ ~

Parker had never been interviewed by network TV or by Big Cable for that matter. This was all very new, and James Cory was big time. Looking into the semi-darkness of the studio, beyond the glaring lights of the set, was intimidating. A makeup tech was blending a second or third layer of anti-glare base and adding a little more black to her already dark eyebrows. "Do you mind if I clip a little bit here, ma'am?"

Parker caught a little bit of Southern in the accent. The tech snipped at an apparently unruly hair. Parker asked, while moving as few facial muscles as possible, "Where are you from? I detect a little bit of Southern girl." She let a little of her own Southernese flavor her question.

"Yes, ma'am. Little ways south of Charlotte, near the border." As she recognized a fellow expat from the South, her lilt became more pronounced.

"Me, I'm from North Florida, which for all intents might as well be South Georgia. Matter of fact, just west of where I'm from they call the area LA for Lower Alabama." She puffed a short laugh "Seems I've moved from LA to LA. We had the same blue-haired church ladies, gingham table cloths, and we were dead center in the heart of Walmart and Friday night football."

"I surely do know what you mean, ma'am. Now hold yourself still while I get these eyebrows balanced."

Her nervous chattiness chastised into silence, Parker became a conscious statue and watched what she could of the studio prep beyond the arched body of the make-up tech. Behind her, where she could not see, a bright green panel took the place of a real set. A book-lined den scene would be dubbed in electronically. In front of her, beyond the clear Plexiglas desktop, three monstrous cameras with teleprompter screens placed in front of the lenses covered most of the possible side and front angles of the set. A roving cameraman was crawling into his Steadicam rig as the host, fresh from his own makeup session joined her. James Cory was respected, thorough, and known for the occasional left field ambush. She took a deep breath.

"Good morning, Ms. Parrish. May I call you Parker? Beautiful name by the way, most singular."

"Well, there is only one of me. I Googled it." She thought she was pretty clever to have come up with it so fast, considering the leaping toads overtaking her gut.

"That's perfect. Keep up that quick wit and we'll have a great interview. You did review the primary line of questions we provided?"

"Yes, I have them right—" She was reaching for a purse before she realized it was back in the green room.

"No, that's not a problem. I just wanted you to know we weren't going to throw in anything you weren't prepared for."

The newscaster looked almost comically orange. Painted shadows were overly obvious. *God, does anyone actually have teeth that white. Probably painted or prosthetics. Too perfect.* "Mr. Cory, I think I'm as prepared for this as any courtroom I've ever entered. I'm ready when you are."

"Fine, just settle in. We should be getting the cue any second now."

"Thirty seconds," came in through the earpiece. Mr. Cory sat back in his chair and assumed the neutral smile of a well-fed cat.

"Ten seconds," then, "and three, two, ..."

"Good evening from PRN. Our guest tonight is the CEO of the planet's newest environmental movement, Ms. Parker Parrish. Ms. Parrish and her partner started up the now seemingly ubiquitous Room for Tomorrow environmental action group just over a year ago. Their original planning sessions in a San Diego County kitchen could not have imagined the successes they've achieved." An insert in the broadcast image showed a crane lifting the RFT initials to the top floor of their building. He turned from camera three to face her, knowing that camera one had his handsome fortyish profile in view and that camera three could get his good left side. "Ms. Parrish, considering that there were already several international movements out there—Green Peace, Earthwatch, the Nature Conservancy, Sierra Club, and a host of others—how do you explain the meteoric rise of your own agency?"

"Well, thank you, but first, let me say that we are not an agency. We do not answer to any one government, any singular cause, or anyone but ourselves and our conscience. I appreciate that the others you've mentioned have done important pioneering work in issue identification, taking the world's eyes to places they'd often rather not look: we've taken a larger view." She shifted in her chair. *Dang those lights are hot. I hope that anti-sweat powder holds up.* She added quickly, "Actually, we've taken a longer view." She thought to herself, *a much longer view.*

"Please do explain." Cory encouraged. Cory's eyes flitted to another monitor and flashed it a signature grin when he saw the red light blink on.

Parker thought, *He is one self-conscious SOB.* She said, "If you take most of the projections that many of the interest groups have compiled and combine their cumulative effects, you might feel as we do that there are crisis points in our collective future that are going to be painful if not deadly for too much of humanity." She had been told to sit still but she found herself wanting to tick off issues on her fingers. "Almost eighty percent of the world's population lives near the coasts of the world oceans. Many of those areas are sensitive to sea level rise of only a meter. With few exceptions, almost every new estimate of the rate of sea level rise

167

has increased the rate of previous studies. So, not only do we have that to consider, we've come to believe that our own estimates are probably outdated by the time we're reading them and will prove insufficient for planning in the coming decades."

He nodded, looking thoughtful.

She continued. "Everyone but the seriously committed deniers understands the principal cause for sea level rise is melting ice sheets due to global warming due to elevated greenhouse gasses. No matter who is creating the gasses.

"We're going to experience more extreme weather, more rain in some areas used to rainy seasons, and longer and more devastating droughts. Global warming is changing climate averages, modifying habitats, and importantly, rendering areas useless which were previously only marginal for agriculture." She thought of blowing sandstorms over the salt-encrusted fields in the upper Nile valley. "Farmland gives way to pasturage; pasturage gives way to barren terrain.

"And then there are regional changes to rainfall." She paused for a breath. "The recent catastrophic flooding in America's heartland, the seeming increase in higher-category hurricanes and typhoons are testament to ocean temperatures which have a higher average by only a few degrees." She took a quick sip from the stage prop coffee cup. "Rainfall that used to supplement the supplies of deep-water aquifers in the southwestern desert communities is dwindling. Ice packs are not as deep, so summer runoff that used to keep reservoirs full is compromised. Conditions we used to insist were due to only periodic El Niño events are now commonplace." This time she paused for effect. "There is a limited supply of geologically ancient underground aquifers, and no one from Palm Springs to Las Vegas to Phoenix seems to see the writing on the wall."

Cory shifted in his seat, preparing for the next question but was cut off as Parker hit her stride.

"That's only here in our own southwest; several million people in the desert southwest will simply not have enough water to drink. This last year, there was a near-record snowpack in the Sierras, and places like Carson City and Vegas will appreciate full reservoirs, but those desert communities keep focusing on growth, not sustainability. When the ground water they've been taking is gone, they will have to finally face the fact that they can't keep on doing the chamber of commerce shuffle to keep building, keep growing."

"You've been mentioning locations across our great country. Looking at Room for Tomorrow's extensive international portfolio of projects, what do you see happening elsewhere?"

One of the massive cameras was rolling closer, and she tried not to look directly at it but kept her focus on Cory's face. "Different areas will see other equally catastrophic changes. When the polar caps finally cease to retain ice through the summer, the additional sunlight captured by that seawater will cause changes in the current flowing between Canada and Greenland. This change is expected to divert the Gulf Stream away from the British Isles and northern Europe. Winters that are already harsh will become longer; and ironically winter snow accumulations will worsen unpredictably. Thawing permafrost across the Canadian and Russian tundras are going to release massive new levels of methane, further accelerating warming and higher rates of sea level rise. These factors are going to force the movement of shifting populations, and you know how unpopular immigrants of any stripe are now."

Cory sighed, perhaps feeling like he was losing control of his interview. He flipped a question card. "Ms. Parrish, with a global economy, won't those regions that are affected by local effects be able to import food, or even water?"

"Maybe Palm Springs can support that expense bill, but not the Third World. Tell a Sudanese cattle owner to import water. Tell a Senegalese farmer not to worry, that the Senegal River will surely come back. All over the world, we're going to lose some of our best farmland to saltwater. Over a hundred square miles of very productive agricultural lands in Bangladesh will disappear under rising tides. Even now, saltwater encroachment from tropical storm surges disrupts their harvest yields for years. When those lands are finally inundated, several hundred thousand farmers with limited skill sets are going to be on the move, or more likely they will end up in refugee camps, dependent on the charity of the world relief organizations."

Cory interrupted again. "Couldn't dikes be built, on the order of the low country in Belgium? Surely, engineers can—"

She cut him off. "Sure, if Bangladesh had the economy of a successful European country. They don't." She wondered why

Cory was becoming argumentative. He was supposed to be one of the more congenial talk show hosts.

"You may have a point there." He shuffled another card to the top of his stack. "So, it would seem that a lot of different factors are going to aggravate each other's environmental impacts."

"Yes, only not so much aggravate as accentuate or add to the impacts felt elsewhere. The main point to grasp is that there are limits to how many human beings a particular environment can support in anything approaching a decent standard of living. Many of the issues we are studying with our foundation grants are in support of methods to mitigate." She paused. *Middle America is watching. Need to dumb it down to seventh grade level.* "Or relieve the impacts of these restricted or eliminated habitats. We are funding electrical battery research, increased photovoltaic efficiencies, and wind farm installations to reduce demand for fossil fuels."

"Isn't that already being done?"

Great, not a challenge question so much as an explanation question. "Of course, in Europe especially." She asked, "Have you ever taken the Chunnel train from Britain to France?"

He paused, confused at the reverse questioning. "No, I've flown that but not the train.

"First thing you see when the Eurostar pops into sunlight in France is a bank of generating windmills." She hid the grin that wanted to acknowledge she'd gotten in one of her points. "Here in the US, the big energy companies are just getting started." He nodded in agreement, but she pushed on before he could get a question in. "But those companies have invested billions in pipelines, extraction sites, and more efficient refineries. Their stockholders are going to want them to use those resources."

"That seems fair." He was about to go on.

Her lips smiled, but her eyes riveted him. "We are not interested in fairness. Whoever invested in Polaroid might have thought that Kodak's cornering of the celluloid film camera market was unfair, just as Kodak's investors thought the digital photographic revolution was unfair." She smiled at him with a questioning raised eyebrow. "I hope you didn't buy stock in 8-track tapes."

He returned the smile. "No. I didn't, I guess you have a Darwinian view of business competition."

"Of course, when the free market is free to respond. As it is, what we see as choices in the marketplace are subverted — corrupted is not too strong a word — by the forces of inertia in the supply chain. Technologies evolve. Remaining petroleum reserves are going to be vastly more valuable and useful in the many forms of plastics and chemicals that we use and reuse rather than a one-way trip through the tailpipe of a Ford pickup."

"You're picking a fight with Ford Motors?"

"Heck no, not in particular. But the American truck buyer seems to have responded to the testosterone driven ad campaigns delivered by all, by every one of the truck manufacturers. I used to be able to step into a Ford F-150. They made a big media push a few years back about aluminum body parts saving weight, and then beefed that model up in size, so you now need a running board to get into it. Senseless. I realize they are the backbone of the average blue collar worker, but the average pickup truck drives around empty." She paused, off track. "But you're letting me rant. What's important is that..." She paused, she had insider information. "Let me rephrase. I know from some of our contacts that the all of the world's major car makers are investing heavily in conversion to all-electrics."

She settled in on what she could say without divulging any NDA confidentiality. "GM and Ford are both fully committed. There are searches now for new properties, in areas hurt by the move away from coal, to develop major new manufacturing facilities, say, Kentucky or Tennessee, where new plants are to be sited specifically to produce the new line of all electric vehicles."

Cory looked as if he'd been scooped but recovered quickly. "That, I suppose is a good thing."

"Of course it's a good thing. We are advocating for the cessation of coal mining and coal burning for all kinds of environmental reasons; these initiatives will provide jobs in areas where there are generations of hard-working people whose lives depended on coal. It's fitting."

Cory checked his note cards and flipped to the next one.

"You have spoons stirring multiple pots. Looking into a wide range of topics of global importance." Camera two picked up on her nodding. "You must have research teams working on that data? We'd like to follow up."

171

"Certainly, but we don't have the research teams. That's the beauty of Room for Tomorrow: we fund the research and political change. If you send us twenty-five or fifty or a hundred dollars, we aren't going to go squirt water cannons at whaleships. We are going to go full press against the Japanese government to shame them into making the harvesting of whales illegal. We'll follow up with advertising campaigns aimed at the Japanese consumer and lobbying campaigns in the Kokkai, their parliament."

Cory was nodding in agreement.

"And why not give those whalers an alternate shot at making money? Money, the J-O-B is always the bottom line. Like the coal miner to car-maker conversion. Let those captains and fleet owners make money cleaning up the world's oceans. Those same whale ship crews can be retrained in a refitted fleet of vessels that deploys to harvest flotsam for cash or that maintains ocean turbines that will generate low-cost and continuous energy."

"Turbines?"

"Absolutely! Ocean current turbines developed for the Japan Current can provide enough kilowatts to shutter all of Japan's nuclear plants for good. We want to provide the seed money for those research and pilot projects. We could do the same in the US with the Gulf Stream and shut down, or at least not have to replace, three to four nuclear plants. Those currents run 24/7, even when the winds don't blow and when the sun goes down. Money for nothing!"

Cory flipped to his last interview card. She realized that he was going to change the subject. She continued. "And the same idea works for the coal producing sector that is railing against change. They don't need to go completely out of business. Emerging carbon research indicates that new carbon fiber reinforced structures are making airliners far lighter, more efficient, resulting in a reduction of the vast quantity of hydrocarbons deposited directly in the upper atmosphere. Carbon research is leading to new types of storage batteries that could be molded into vehicle body parts, so the photovoltaic roof and trunk could directly charge the carbon-based storage liner immediately below it, all of it being held up by a carbon fiber reinforced chassis. How cool would that be?"

He raised eyebrows as if he was impressed but quickly changed the subject. He said, "Ms. Parker, you currently occupy most of a small six-story building. How do you plan to take on such a tall order as, as, well, saving the earth?"

"That's a very good question, Mr. Cory. We're expanding rapidly and will soon fill our existing site. We're planning an expansion campus, possibly over in the Valley a few years from now. Most of our space is now occupied by phone bankers, contract attorneys, think tank war rooms, and we're a training crew for grant managers. We're running out of room." She turned toward the camera that had her in side view, seeing that the red "on" light stayed lit. "We're looking for a few more good people right now." The wink sealed the deal.

~ ~ ~

As he watched the tape delayed broadcast from his office Carl thought the interview went perfectly. Parker disarmed James Cory's interrogative style and projected the perfect friendly witness for the defense. An alarm went off on his watch. He tapped the watch screen and scrolled to the icon for his schedule. A red dot blinked on 5:15. Damn it! He hurried downstairs to join a large crowd that was heading for the parking lot. An erector crane's diesel could be heard from beyond the row of low trees lining the lot entrance. The building's former bank logo had been removed earlier and replaced by the RFT letters.

Amid the applause from his employees, a sixteen-foot diameter LED image of the Earth rose to join the ten-foot high RFT initials. The simple sign would show continents in slow rotation— so slow that in the next few weeks, many of the building's new employees would go into the parking lot on morning and afternoon breaks to be convinced that yes, the logo really did imitate rotation. The image currently displayed represented the side of the Earth facing the sun. The signage was the brainchild of a former Disney animator now on their graphics production staff. Looking up at the installation, Carl thought, "That is truly something." In the back of his mind though, he wondered about the darker secrets he was hiding. The secret funding that could bring down Room for Tomorrow.

PART 2

Bruce Ballister

19

The inner sanctum of Abadon Industries was sumptuously spare: clean lines, elegantly simple furniture, and expensive art. It all looked like money, well and freely spent.

"You're absolutely sure? Bet your life on it?" Richard Abadon looked crisp in his gray linen suit despite the late hour.

"Yes, sir!" In comparison, Odie Peck was a character study in self-indulgent comfort. His shirt was clean but not pressed, his khakis a little too shiny, their permanent press creases long gone. He'd never gotten used to his new pay scale or the global status of his current employer. "Not sure about betting my life in any kind of bargain, but I'm pretty sure I know who's been inside." He smiled in the knowledge that he had tracked down an advanced hacker and set his system up to recognize the intruder.

Peck reviewed his mental bullet list. "I've traced his IP through seventeen jumps. There were three false cut-aways, but the end of the trail keeps ending on the same few machines. Those machines move around from server to server, but it's the same batch of servers. I've got him." Two of those cut-aways were in Manchurian-flavored Cantonese and he had still found the bastard. The end of the trail showed up sited in Xinzhou, home of the Green Army Corps and the Crab Group, two of China's best industrial hacker groups. He tried not to grin.

"Laptop?" The junior Abadon raised one eyebrow in practiced skepticism. Richard was the younger of the two industrialist brothers. He no longer exhibited the brash, go for the throat, buyout strategies that had made him both feared and respected in the back rooms of the Chicago exchanges. But he was far less genteel than the six-years-older brother. Richard had gone to Northwestern, older Andrew to Harvard's business school.

Andrew was less likely to be seen on the factory floors or their numerous and varied acquisitions, favoring closing deals on the back nine or the fantail of one of his three yachts. Andrew's urbanity and Richard's more earthy pragmatism often provided different avenues to a deal, but both closed. Richard found the challenge of the missing funds anathema to his very being. Someone was actually stealing from him. Not that he was above zeroing out pensions or shuttering factories on a Sunday morning. This was different; someone was actually siphoning *his* money.

Odie appreciated the insight Richard showed in his instant conclusion. "Yes, sir."

"Well, who is it then? Save me all the geek speak. Who's been taking my money?"

"Sir, these people steal identities as easily as they steal old ladies' savings accounts. Based on his entries, I don't really believe he's a graduate of the Green Army Corps studying at UCLA." He paused. "There's another possibility too. One of the trails I've been following leads to an environmental group in California. It's called Room for Tomorrow. Been growing really fast apparently."

"I've heard of them." Richard turned toward the rooftop garden that fronted his and Andrew's offices. His brother was walking among the potted plants on a cell phone. He turned back to Peck. "What makes them a likely source? Bunch of damn tree huggers."

"One of their senior partners used to be in the cyber security business. He's got skills."

"Well, if it pans out, just be fucking discreet. Be absolutely sure before anything goes down. Got that? Absolutely fucking sure. Let me know before —" He stopped in mid-thought. "No, on second thought, when you're sure, and you are very sure, I don't want to know a damn thing about it."

"Right, I'd rather wait until I can ensure that the certificate ID I've got is the actual person doing the mischief."

"Mischief?" Richard Abadon's face distorted in anger. A single blood vessel in the center of his forehead bulged as his blood pressure rose. "Mischief? Some son-of-a-bitch is stealing millions of my money. That's not mischief. That's somewhere out there far beyond mischief; Hell, it's far beyond the concept of grand larceny. You could support small countries with the money he's stealing."

A meek, "Yes, sir" was all Odie could get out. He wished he hadn't been clasping his hands. It looked sheepish, and he had come in proud of his work. "But, Sir —"

"Sir shit! I want to know who that motherfucker is, and I want to know soon." On his turf and in Italian tailored business dress, Richard Abadon in full fury was awesome, terrifying. "Do you have any idea how much is missing?" The question had been rhetorical. "Fucking millions!" Peck opened his mouth to respond but another yes sir seemed ill-advised. Richard continued. "I'm going to authorize a quarter million into an account that will pay for professional PI services. I want you to ID, track, and find this little shit. I want to know WHO it is. Am I clear? And when you come back, there won't be any oatmealy, "rather not say" bullshit. We understand each other?"

This time, the yes sir covered it, he only wished it hadn't sounded so much like the submissive yes sir. His stepfather had always insisted on it, and he had always hated his stepfather. "Mr. Abadon? There's one thing."

Richard fixed him to the floor with a laser stare. "What?"

"These Chinese, if it's them. What happens if it's one of the Triads?"

Abadon snorted. "Chinese Triads?" A flicker passed behind his eyes. "If we get to that point, I'll make some calls." He extended a hand toward his office door, indicating Peck's dismissal.

"Yes, sir." Peck turned to leave and as he reached the door…

"Peck."

Peck turned, he had just opened the door and pushed it quietly closed again.

Richard Abadon's mouth was a thin, mean line. "Refer to this in any communications as 'Weed Cutter,' understand?"

"Yes sir."

"No, Peck. I mean, do you understand; Weed Cutter?"

"Oh." Peck gulped down the fatal implication. "Yes, sir."

He passed through the outer office and nodded a smile toward the silver-haired receptionist. The return smile fueled a fantasy that took him halfway to the ground floor. *Weed Cutter. Shit, better call my old buddy Tom in on this.*

~ ~ ~

The trip to the CalStation mountain house from Anaheim often took well over three hours in drive time; an hour and a half longer than their usual Sunday morning check in. This Sunday morning was sixteen months after Parker's first encounter with the truth of the room in her 'dream' house. Grieg had said there was a surprise for them, but he had not shared further. When Carl said he'd follow her out later she was at first disappointed. But going alone had its benefits. She'd take the Jeep over the mountain, her first route to the house, which was a lot more fun than the easier route around the mountain.

This time, her mood was not dampened by the sudden appearance of clouds. The morning was clear and bright thanks to a high-pressure ridge over the Sierras that was pulling rain-rinsed clean air down over Utah and Nevada. You could almost forget the levels of smog that often choked the valley when you could see the lines of mountains on the northern horizon fading in shades of dusty lavender.

After the third visit, she and Carl had been parking behind the house, out of sight of the rare lost traffic that found its way onto the usually closed road. They'd thumbtacked a "No Trespassing" sign to the front door to help discourage any random foot traffic. She took the kitchen entrance with no residual feelings of dread. The only thought she'd had was how routine their visits had become since the Sunday that had been remembered as the yester-today. Grieg had given her the excellent advice to hold her breath and tense her rib cage as she stepped through. This Sunday morning, she did just that.

Chu looked up from his usual workstation and smiled in welcome. "Good evening."

"Hi Chu, good morning. How's your project?"

"Near finished."

"Great!"

"Maybe not so great." His upbeat greeting had faded.

"What's wrong?" She sat at the desk station, twirled in the seat to face him. "What's the problem?"

"Nearly finished is the problem." He chewed absently at a nail on his left hand. "Breaks are happening. Breaks in the timeline."

"Breaks? I thought you said time paradox problems were avoidable."

"That is true; they are avoidable. But we have not been obeying the protocols." He put his left hand in his right, as if that would damp the nervous energy that demanded nails be trimmed by tooth. "We have been doing far too much that's forbidden in the protocols." She noted that his English pronunciations were much better, the dis and dat replaced, finally with this and that.

"Is there evidence on your side that we are making a difference?"

"It seems that way. Yes. I often make inquiries to the genealogy banks. People are missing."

"How can that be?"

"They were never born. People, friends, cease to exist. Or, I guess I should say, people that I have known have never been born; their parents never met."

"Oh, I see." And she did see. Something, one of the interruptions, or interventions, nearly 200 years before Chu's present was affecting outcomes down the timeline.

"Are you in danger?"

"I'm not really sure, Parker. In training, we were told that a large enough change could disrupt the family tree above us, that we might suddenly go poof!" He flicked his fingers in the air as he might fling off droplets. "I don't know if that's entirely true if I'm on this side. Could I engineer my own disappearance? What if I step through the barrier into nothingness? We didn't invent this technology." He waved his hand at the room in general and his console.

"Who did? Space aliens? Who else is there?"

"What I mean is, humans did, before the war. It's from the 2120s. Friedrick was working on designs for an electronic shield, an improved Faraday cage that could withstand directional electronic espionage or an EMP wave front. His power feed got hit by lightning and he and his cage hopped fourteen and a half minutes into the future."

For almost fifteen minutes his entire apparatus was non-existent, his crew were gwin'on about 'hoppin on the eternity train.' She reacted to his cultural drop with raised eyebrows. "Sorry, lyrics from a perf piece that was popular then. Then Friedrick was right back. Lucky as damn hell that no one was standing on the spot. Ian was completely unaware."

"Ian? Who was Ian?"

"Ian Friedrick. The tinkerer. He took the first hop. Maybe I'll take the last. I don't know, Parker. Maybe I've taken the last."

Parker considered how much Chu had just revealed. This port, or gate, was not the early 23rd century's creation. It was almost eighty years old. "Chu, when did time traveling start? Why?"

"Friedrick's first hop: that was 2112. The first news story broke in 2115, instant negative reactions in the media circus. Lots of debate, lots of hate. Resolutions in the UN. It was eventually revealed, in 2118 I think, that he had been doing it for a while. A few ports were set up around the country at first to test the process, a few more set up by universities in Berlin, Milan, Beijing, Cape Town. Others were under construction when the war broke out."

"Why here? Nowhere California?"

"Well, remember at the time, it wasn't nowhere. That city down there was a major tech innovation center with a collection of gleaming state of art architectural towers and residential arcologies. And it's close, reasonably close, to one hell of a lot of American history and culture. Any closer to the coast and the gate would have been more earthquake prone. We're almost sitting on the damn fault line, but the rock here has smoother, long wave seismic properties. Other side of the fault line, the gate might not have survived."

She saw the true nature of the station's sturdiness. "I..." She faltered. "So, this station was here when—"

"Yeah, that's why the damn stairs are so ricky-ticky." He started to nibble a finger, stopped. "I think, too, this side of the mountain is damn near quake proof, even though the fault is only half a click up the hill. Most of the shakin' and breakin' goes on west of the fault."

"They'd know, right? I mean. There'd have been seismic records going back centuries, right?"

"That's good! Fast thinking."

"But were they any good at transporting into the future?"

"That was in the planning stages. If we could use this room to get to the future, to a time when the outside was no longer toxic. Say another 100 years forward."

"Yeah. Maybe you could go out and find growing things had returned."

"Technically," he paused, looking down toward a few small drawers built into the cabinet below the countertop. "This station is one of the two that were being fitted for forward travel, but no one has ever done that. It was used to study Americana. The morality of it was being debated by the same groups that had been trying to control trips into the past. Anyone from the mid-22nd was long dead by the time we found records of the gates and decided to make initial forays back. We need technology that's been almost forgotten."

"You said the time slip was constant. What were they studying?"

"Well that was back just before the Last Day, so they were looking into the rise of the American Super Culture that arose after the second world war."

"Hm, Super Culture."

But Chu's attention was back on his work. He started flicking his fingers at his terminal. He paused, looked down at the control panel, and then back up at Parker. He was about to respond when a soft buzzer announced someone in the outer door. Parker turned in her chair to make room in the narrow passage. She had just moved her legs out of the way when the gate flickered, and Carl made a clumsy entrance through the portal. She said, "Hey Grace, dance much?"

"Ow!" Carl kicked a stool and wobbled a bit trying to regain his balance. "Dammit!" His momentum carried him forward, and he fell forward onto Parker. They soon recovered, laughing. "Damn Chu, couldn't you guys have designed this thing to have, like maybe, a soft-landing zone on this side of the gate? I mean, we have to almost jump through, and then there's furniture on the other side."

In a few minutes they'd brought Carl up to speed. Chu's generation had nothing to do with the design of the port. The conversation was interrupted when the proximity buzzer triggered again. All three heads swiveled to the source. Parker asked, "Is Grieg out? I thought he was back in the 23rd."

"No, I mean yes, he—" Chu stopped, stood, backing away from the gate. He waved his arms in alarm and whispered just above hearing but with a force that implied command. "Grieg is out; he's out there." He glanced over to the outer doorway to the decontamination chamber in a combination of meaning and confusion. "Get up, clear the way, NOW!" They did.

All of them moved to the second room. If there was someone in the house, in the 21st, there was a chance that they might blunder into the front room. Before this visit, neither Carl nor Parker had seen any sign of a weapon in CalStation but now, Chu was holding one.

20

Grieg steered his steel-wheeled ATV around one of the many switchbacks that threatened to kill him each time he drove back up to the CalStation time gate from the valley. His personal pouch held three full data cubes that held real promise of allowing the chem-techs back in the NZ a chance of rebooting a gene-splicing pharma industry and curing at least half of the most common exposure related cancers.

Although hot and dusty, the drive from Neenach had been uneventful. It had given him time to think about his past. Griegor Grigoryev was inducted into the Niners just after the New Year's celebration of 2200. He had hoped that his father would attend, that he would have survived until the new century, but his father had been too sick to travel. His mother had died of acute lymphocytic leukemia twelve years earlier, before Grieg graduated out of lower division. Young Griegor had taken history as his prime line in upper division and become an expert in Balkan politics in the period leading up to the third Jihadist uprising. His secondary line had been cloud integration. His secondary line study group, Cloud Nine, selected its members from the top five students in each of the four years of upper division study. Cloud Nine, self-named the Niners, was essential to the global reboot; its graduates were dispersed among the United Colonies as communications and data retrieval experts.

Niners were responsible for maintaining communication with the few remaining orbital satellites and keeping open the remaining lines of communication between the far-flung elements of the UC. They had reprogrammed the satellites to receive uploaded data and only download in bursts when they were approaching a target's airspace. The early Cloud depended heavily on both undersea cabling and the constant array of orbital geosynchronous links. The Cloud was first developed as pure storage for vast amounts of offline data storage. Its secondary incarnation was the interconnected halo of communications satellites. In the second Jihadist uprising, fundamentalists had attempted to disable the halo, fearing that it might become self-aware and thence god-like.

They had managed to take out about a third of orbital memory capacity with EMP burst munitions before their launch sites were bombed to dust motes. The solar flare of 2192 had taken out most of what was left. The geosynchronous satellites serving NZ had been fried in the later stages of the Last Day. The few shielded Commsats still aloft had limited orbital lifetimes remaining. The first age of satellite communications was near an end in the early 23rd century.

By the time Grieg graduated at twenty, he knew almost as much as anyone in his time about the lost computing sciences of the terminal generation. His parents, children of the terminal generation, were called the seminal generation. For in that generation's germ bank lay any hope of survival for the species. Grieg was second gen. Second-geners had heavy pressure to rebuild, to lay foundations for a new society that was not shackled by overpopulation and ethnic nationalism.

There was still poverty and misery in abundance. The second-geners had significantly fewer cases of mental illness; they had been raised in an atmosphere of hope rather than despair. Grieg's assignment at the dangerous California gate was extraction of the last bits of useful technology from one of the world's few surviving cloud storage sites.

His work in the CalStation was almost complete. He'd extracted as much as he could on genomics, metallic plastics, and transform mathematics, and now finally cold-banked fusion. Distracted, he approached the cabin on his four-wheeled, metal-tired electric ATV and drove to the small inflate-a-shed they used for a garage. He hit the brakes hard. There was a new addition to the time gate's backyard.

Staring gap-jawed at a rusted hulk at the back of the farmhouse, he had to wonder what just changed. The weathered remains of an early 21st sedan rested on its rims in the low scrub that was trying to survive in the harsh post-human habitat of

Southern California. Why now? Something had happened in the 21st while he was at the Chick-a-Saw facility. "Crikey!" he said aloud through the muffling filters of his rebreather. If there are visitors in the gate, they could be law enforcement, paramilitary types from the corporate cops of the 21st or almost any third-party tracking Parker or Carl. It didn't look like a police vehicle, although it could be park police. No paint or decals remained on the sand-blasted hulk.

He sat in the driver's seat, squinting through a rain spattered window. It wasn't likely that a third-party friend option had occurred. Someone had arrived back in the 21st and not left. *If that don't put teeth in the shark!* There was no way to get into the gate with stealth. The decon room was plasti-paneled so team members could see if anyone coming through was in trouble with air poisoning. "Nothing for it but the try!" He stepped over the remains of the farmhouse foundations and climbed the ladder to the front of the elevated time gate.

Grieg stepped into the outer door as quietly as he could. Through the plasti-panel he could see four people staring back at him. The one he didn't know had a gun on the three he did know. "Crikey!" He toggled the small audio button to "live" in the room. He kept the breather on and removed everything but his undergarment. A cloud of hydrogen flushed most of the gasses from the room and then normalized with a blast of clean filtered nitrox. He could hear yelling in the room outside the chamber, but it was muffled by the rushing gasses. As the partial pressures stabilized and the hissing stopped, he sensed the ongoing conversation was intent on keeping him in this chamber.

He thought, "*OK, going to put on a show.*" He looked quizzically at the gesturing 21st-ers, and the gunman. He put a hand to his ears, gesturing back that he couldn't hear. He made an obvious point of flicking the switch for exterior sound off and on and silently mouthing a shout to prove he couldn't hear that he was being ordered to stay put.

~ ~ ~

"What do you mean he can't hear me?" The gunman had every reason to expect that the man coming in through some kind of airlock was faking deafness or at least audio equipment failure. "Tell him to keep his ass in there."

Chu moved two steps toward the translucent half-panel fronting the decon chamber and put hands up signaling Grieg to stay put; one bandaged hand showed blood spots. As he lowered the hand, he flashed the universal computer hand signal for silence — two horizontal fingers, knuckles out signaling secure connection. This maneuver accomplished three things — it separated him physically from the locals; it let Grieg know the visitor was dangerous; and it put him closer to the little cabinet that held at least three weapons; a projectile handgun and two spritzers. A spritzer was normally nonlethal. Its two needle cartridges were charged with opposite polarities. When they landed millimeters apart in their fleshy target, they discharged a high energy spark that never failed to be its victim's new center of attention. Occasionally, a head or neck shot was fatal. Chu was in no mood to maintain protocols about damage to citizens from the 21st. His hand hurt like hell. The effing slag had shot him in the hand.

Chu turned to survey the situation at the entrance. Parker was in partial shock. The gunshot that stopped Chu's charge with a bullet through his hand had frozen her into stunned statuary, but that was quickly morphing into anger. Carl was seated at the small table, hands up but poised. He actually appeared to be coiled like an over-tightened spring. His jaw muscles rippled with anxious tension. The room had just stopped reverberating from the gunshot when Grieg began to cycle through the sterilization chamber. The gunman was staring wildly around the small chamber.

Parker found her voice. "I don't know who the hell you are, or who sent you, but you have no idea where the hell you are, and you'd better back off a little and find out what you've stumbled into!"

The intruder turned to her. "Well, I'm Thomas C. Corliss, at your service, Miss Parker Parrish. I sure know who the hell you are, and you'd better hope you live through the next ten minutes. My boss has a burning desire to know who's paying you, or rather how you're stealing from him, but he'd be just as happy to see proof that you're a corpse. Really, lady. I don't care. You're mine, dead or alive." With his quarry cornered in what used to be a farmhouse, he took time to consider. "What the hell is this place?"

"Does it really matter?" Chu said. "Spaceship, time machine, teleportation chamber?" He put out two hands, palms out, gesturing a polite calm down. "Why don't we just slow down and talk a bit. I'm sure we can figure something out—glass of water, maybe?" He took a short tentative step toward the intruder. "Agreed? Can we calm things down? We can talk about how the farmhouse turned into a laboratory." He tried a smile, "Please?"

The gunman's eyes were darting around the near and far corners of the CalStation, back to the faces and hands of three current occupants, and back to the slowly cycling chamber behind them. His pistol's aim seemed to move with his focal point.

"Teleportation chamber?" Corliss's brows furrowed as if finally registering what he'd heard with what he saw.

Carl started to move toward him. Corliss noticed and swiveled to his right to cover him. Parker made a move to her left and forward a half step. A wave of the pistol toward her stopped further progress. Corliss glanced quickly behind him. The opening to the farmhouse still shone as the energy field played on dust motes and circulation currents. He couldn't back any further. Corliss tried to maintain control of a situation, and all could see the confusion on his face. After all, he was siding away from a shimmering panel that had been a door only moments before. Too much was going on at once.

Over the hiss and clicks of the machinery in the detox chamber, the four occupants of the main work area eyed each other for advantage. The intruder, scanning the three closest to him, reacted to another movement in the detox chamber. "Everyone stay the hell put or I will not be aiming for hands." He took stock again, looking for the other one. The one in the little room was gone. Where was he? "Hey, you? In the other room, show yourself."

The lights went out. Parker and Carl both moved immediately toward the gunmen. The gunman fired. Carl yelled out in pain but did not stop. The three went down in a heap. In the dim glow of panel indicator lights, Chu could see that the hand with a gun was vertical and restrained by both Carl and Parker. He took quick aim with the spritzer and shot at the gun hand. Another piercing yell filled the room as Corliss's pistol flew away. The intruder's wrist had blossomed in an arcing spark, and the smell of burned flesh filled the small space.

~ ~ ~

Carl sat in an overstuffed white channel back chair in Parker's living room. The drive down the mountain had been difficult but the 23rd century drugs helped with the bullet's damage to his side. They'd decided to forego a falsified police report of a close call mugging and let the yellow goo do its work. Neither really wanted to involve the police in a who, what, where, why, when line of questioning. Chu claimed that the yellow gel contained antibiotics as well as encapsulated stem cell globules that would phase in as the antibiotic effect diminished. The degradable staples were good for four weeks, after which Carl could carry on with due caution. He'd bravely said, once the gel softened the pain. "It's only a flesh wound."

Parker grinned with hint of malicious pleasure. "Monty Python? 1970 something?"

He resented house imprisonment; Parker wouldn't let him go to the office wounded. But it didn't keep him from his keyboard work. It was a new machine, set, as usual, behind double blinds and cut-aways. When he was wired, a ping in his direction would show Fort McMurray, Alberta, or the Daxing suburb of Beijing. On Wi-Fi, he only compromised his connection when he was in his own search and destroy mode. As soon as he'd closed in on a target and downloaded any results, he powered down, opening the air gap, and disappeared. He'd missed getting closure on Corliss. He hadn't been conscious when the partially repaired assassin had been turned loose in the 23rd century. But he wasn't disappointed when he learned Corliss wasn't likely to survive forty-eight hours outside without a breather.

Parker had relayed Grieg and Chu's predictions for Corliss. That Corliss would discover his car suffering from almost two centuries of neglect. The confusion he'd have when he examined the outer shell of the elevated time port, it's fabri-steel supports adjacent to the farmhouse's stone foundation. His further confusion when he found the uninhabited ruin of a city in the valley. Even if he had hand print authority for a three-wheeler, its range would not get him to the highly radioactive coast. He would remember that his only chance of survival would be to the east. But he would not have known about the renegades or the canines.

Considering Corliss's fate was depressing, so he began to appreciate the endless line of swells moving toward shore. Carl's view of the Pacific was a changeable tableau, offering a different view every few minutes as low clouds of varied darkness scudded overhead. He heard the soft mumble of a car entering the garage and shortly, keys in the door. Then, bags and the shutting of the garage door. "Out here, on the veranda!"

"Oooh, the verandah?" Parker affected the Southern accent of her North Florida pinelands youth. "Shall I bring some juleps, with mint?"

"I'll take Scotch, the older the better."

He heard, in order, the tinkling of ice cubes and glass, a gurgling of liquid, high heels being kicked loose, and then her soft padding three steps down to the lower level. Parker's glassed ocean room offered views up and down each side of their cliff and the gray-blue outlines of offshore islands. Just beyond the lower level's glass wall, the outer porch and pool deck were cantilevered out over the cliff face. It was spectacular and one of the last built before updated earthquake codes forbade their construction.

"Hey gorgeous."

"Hey handsome." She handed him his drink. He studied her feline curl into the matching chair. She swiveled to look out at the darkening evening cloud deck. "No sunset tonight."

"I never give up until it's dark. Beyond hope."

"Never give up. There's a motto worth living by."

Carl couldn't stand not knowing any longer. "How'd it go with the boat?"

She raised her glass and pointed toward a seascape at dusk; only white caps and surf were visible beyond the salt-stained window. "The *Sails Call* is in deep water by now." She took a sip. "I had Jeremy and Ahmed go to Corliss's marina. Convenient that he had boat keys and marina receipts in the glove box. Rent one, steal one. Jeremy had the navy-blue hoodie and jeans. He was the closest to Corliss's height." With full tanks and spare fuel in jerry cans, they took both boats out into the morning fog. They'd concluded before they actually scuttled the *Sails Call* that any remaining fuel should be transferred so that the rental would actually make it back. She shortened the story considerably. "The Catalina Basin is about 3,500 feet deep. The *Sails Call* has hooked its last fish. Tom Corliss will be presumed lost at sea."

191

Carl thought about that a minute, aware that Parker was staring. He pivoted away from the darkening western sky. The amazing painting of the kelp forest's depths got him thinking about Tom Corliss. What about Tom? He took another sip of Scotch, letting the sharp bite of the liquid pause and deaden his tongue to any other sensation. He realized he had thought of the unfortunate hit man as Tom, not Corliss. He had no imagination to fill in the blanks. He let the Scotch go down and closed his eyes.

He promised himself to go through the back door, to help if he could on one of the missions to the valley, or at the very least, to step out the back door. It would be good to actually know what had happened.

~ ~ ~

Odie Peck placed his visitor's key card against the sensor and pushed the elevator button for 8F. 8R would take an employee to the bank of executive and administrative offices for those whose rank in the organization required almost immediate and proximate access to the top. The occupants of the larger walnut-lined offices were also major shareholders who had major personal interests in maintaining that portfolio. 8F served only the two executive offices and required key card entry approval at the lobby level. The doors opened with a soft ping. Peck stepped into the reception area for the Abadon brothers and their executive assistants. He stepped forward and veered right toward Richard Abadon's reception area. From the broad carpeted entry, he could see through the glass-walled conference room separating the two brothers' offices out to a rooftop garden. Hell, he thought, with the wealth these guys had, it might as well be the Hanging Gardens of Babylon; this was subtle, understated wealth.

He thought back to the last interview with Richard Abadon. It had not gone well. He hoped his news would be more welcome, even if it did have several loose ends.

"Mr. Peck, please have a seat. Mr. Abadon is delayed in a phone conference and will see you as soon as he can." He took a longer appraising look at Richard and Andrew's receptionist. She was possibly a little over fifty, only a few pounds off her college graduation weight, and almost certainly a Kansas native from her plains accent. Her silver hair seemed all the more attractive with the temple hair tucked behind ears and into sweeping lines that ended in a short ponytail. *She makes getting older look sexy.* As much as he wished otherwise, at no time did her demeanor and delivery break from a cordial but professional wall. "That," he thought, "was why the Abadons kept her minding the door."

He took a seat in a leather-skinned chair that was more comfortable than it looked and prepared his thoughts for the interview. His hunch that the one of the data trails led to Carl Reyes was paying off. His electronic surveillance team had searched for any pattern in the stream of donations that funded the new and very public Room for Tomorrow organization. Reviewing his report, donations were strongest from college towns and large urban areas, and there was significant support from areas that had major federal or state park lands. They were lowest from oil- or coal-producing areas, but this was no surprise. There were large single donation sources across Europe. He had tried to track down some of the large single donor sources, and this is where he'd found inconsistencies. Some of those were from foundations that supported both Room for Tomorrow and the plethora of other eco-groups, including the more aggressive Greenpeace.

Peck had investigated these ties to the ends of their discoverable data trails. They had been covered artfully and deviously well. Some of these donor foundations were found to be spurious, if not fabrications. Addresses led to vacant lots or vacant floors of office buildings. One foundation led to a Frankfurt pawn shop, another to a bakery in Aberdeen, Scotland. His main conclusion, when he'd fact-checked the initial findings, was that the real money behind Room for Tomorrow was going to great lengths to hide behind false fronts. He could not understand why the more clandestine big-dollar donors would be false fronting their donations to Room for Tomorrow, which ostensibly was working public opinion and, more importantly, promoting research in promising alternative energy technologies. He knew that in a very few instances the Abadons had also invested in some of these emerging techs. Why show your donations from a Scottish bakery?

One pattern was puzzling. Wouldn't these donors want tax credit for some five- and six-digit donations? He hadn't been able to get past the IRS firewalls to see if donations to either organization were being used for tax credit. His statistical analysis of the pattern of donation to RFT, as he'd come to think of it, showed most of them were very typically twenty-five- to five-hundred-dollar tax deductible donations coming in through web crawler and direct mail appeals. The largest ones, the six-figure and up donations, came from false front donors. He had come to a personal conclusion that those largest donors to RFT were the same large donors feeding money into the Verts Sans Frontiers account he'd found at the Banca d'Italia branch in the Port of Gioia Tauro.

The Italian port city had a long history of smuggling, espionage, and human trafficking going back to the Middle Ages. If there was any city in Italy most likely to be involved in quasi-legal or illegal international finance, it would be Gioia Tauro. Financing of an eco-terrorist group, or any terrorist group, wouldn't surprise anyone. It made sense, in a perverted way, that the Verts Sans Frontières would hide in plain sight in a private Banca d'Italia account labeled simply VSF Enterprises. What troubled him was the admittedly loose connections between VSF and RFT.

Were the large money donors to the Verts hiding their donations to RFT? In very general terms, they had common cause but certainly vastly different operating methods. Wealthy backers of the green movement would certainly want to distance themselves from the Verts, but why the RFT? Was there any reason to hold back on making the statement that there was a connection, and the connection was money? Sitting there, staring out at the potted plants on the garden terrace, he realized it was the timing. If anything else connected the two groups, in his mind, it was the coincident rise and successful fundraising of RFT and the rise of international incidents attributed to the Verts.

Peck also had to report on the disappearance of Tom Corliss. No report, no new anything. He'd simply said, "I've got a lead on something big; I'll call in the morning." Nothing.

He began to hear raised voices from one of the two offices. He had shifted most of his attention to listening in to catch any decipherable words. *Good God, someone in there is catching hell, and I'm next.* He heard a door slam, and the voices stopped.

"Excuse me, Mr. Peck?"

He turned, startled. She was leaning over the back of his chair, closer than he'd realized. He'd been caught eavesdropping. "Mr. Peck, you can go in now."

21

Peck soon learned that not everything was a picture of corporate grandeur on the eighth floor. Special Projects Consultant, Odie Peck, squirmed in his chair. Richard Abadon had him pinned to the seat. Sitting beside him, Abadon's Director of IT Security, Harry MacCauley wasn't any happier.

"Let me be sure that I understand you. Weed Cutter has failed. The man you hired to destabilize Room for Tomorrow has disappeared?" Abadon's face betrayed suppressed incredulity.

"Yes, sir." Peck licked his dry lips to continue. He thought, destabilize was a nicely sanitized word for murder, or even decapitate; he'd remember it. "It appears that the contractor pursued some activity in San Diego, Orange, and Los Angeles Counties. His cell phone tracking is all over the area, from the RFT offices in Anaheim to the two principals' home down near La Jolla, to occasional trips North into the mountains or to the East across the desert. He was on the trail of that latest missing draft of 243 million. But his reports are unbelievably circumspect, even over secure lines."

"Circumspect? He didn't file adequate reports?"

MacCauley, Director of IT Security, offered, "Sir, thinking about it, I believe he fancied himself as a combination spy slash hit-man. I think the espionage angle went to his head." Peck decided to add. "We believe, too, that there is an element of PTSD in the profile we ran before we hired him."

"He was unstable then — unpredictable."

Peck thought that maybe Richard Abadon will see this as an unfortunate setback, nothing more. "That's how it turned out, sir. We took his PTSD claim with a grain of salt because so many vets are seeking disability payments. And, if true, it might have led to his new line of work."

"That being a gun for hire? Aren't you also a — ?"

"Sir, the usual word is mercenary." MacCauley painted lipstick on the pig.

There was a brief silence in the room. Abadon swiveled around in his chair, his back to the two investigators, apparently scanning the materials in the dossier again. Time enough for private eye, investigator, mercenary, or simply hired gun Odie Peck to briefly take in the details of the opulent office. The room was nearly as large as his mother's house, held one large desk, two comfortable couches that faced a long coffee table, and a wall of closed cabinets that had begun to work on his imagination when he was brought back to the present bad news. A corner glass wall looked out over the roof garden he shared with his older brother.

"He was last seen in San Pedro?" Abadon had turned back to face the two.

"We have marina cam shots of him, er, someone obscured by a hooded sweatshirt, loading about ten jerry cans of fuel onto his boat and tanking up at the marina. He left with some additional range."

Abadon waived the thin blue dossier at them. "This says it was foggy that day. That would have prevented any aerial surveillance. You're telling me I've got 243 million driving out into the fog?"

Peck swallowed hard. "That's true, and the radar station at the south end of Catalina Island doesn't keep a taped radar record of movements in the channel for more than 24 hours. Weather? They go back a week. Marine traffic, zippo."

Harry MacCauley pivoted to Peck. "Mexico?" The thought hung pregnant. "He could have sailed almost anywhere down the west coast of Baja."

"Maybe so, Harry. But the *Sails Call* is a power boat. I think Corliss probably could have made it with the extra fuel. It's about 150 miles from Costa Mesa down to Salina or just a bit more to Enseneda. Those are the nearest Mexican ports and not much of a stretch for a small power boat. One with extra fuel aboard."

Richard Abadon looked at his watch; the man had plenty of other things to do on any given day. His eyebrows pinched together; lips pushed out in a frown. His piercing stare bore into Peck. "You're sure you had the correct boat? I see the registration for *Sails Call* to Thomas Corliss."

"Yes, if you look, you'll see that it was a Bertram 32. Those are power boats. In the video we have from the marina, the hull had a large profile of a sailfish painted on it. Hence the name. And the time stamp verifies its time of departure."

Abadon thought about his own boat, a thirty-six foot sloop that would need at least one other to assist in sail handling. Thirty-two feet seemed entirely too small to go out into the Pacific alone.

"I see." Abadon shut the folder and handed it across the desk. Peck had to lean across the desk to reach it. A thin smile played briefly across Abadon's lips. "Well, there is a bright side."

"Sir?" Both men said as one.

"We won't have to pay him the second 100,000." Richard Abadon leaned back, glowing in a small victory for his side. As soon as he'd said it out loud, he realized that was a pitiful addition to the balance sheet compared to the missing 243 million. His brother was going to be livid.

"Yes, sir." Peck was sliding the dossier into his cheap attaché case. "One thing to think about, sir."

"And what's that Mr. Peck?"

"The cyberthief may be female." He instantly reconsidered. He'd been wondering about if for days. "Or maybe just a cover."

"What are you getting at?"

"Sir, one of the identities I've traced a couple of times. She, or he, is very good."

"Yes?"

Peck was a little proud of having followed the trace almost to a physical location, he just had to share. "The identity closest to some of the illegal activity is known as FrICanD." He spelled out the tag with the caps. "Get it? Eye candy? A chick that thinks she's hot?"

MacCauley rubbed his chin in thought and looked up. "And why might that not also describe a buffed-up gym rat, who also happens to be an expert Darknet guru? One who seems to be running you in circles."

Inwardly, Peck deflated. He hoped it didn't show. Best to move on. Ignoring MacCauley, he addressed Richard Abadon. "Is there some additional action, sir, on Weed Cutter?"

"First, just to make sure I have this straight. Your guy, this Corliss guy. You say he texted you with a message that he was on to something hot and the next thing we see is him loading a boat with extra fuel?"

"Yessir."

"And you don't have a clue as to what 'something hot' is?"

"No sir, that's correct. He—"

"That's almost absurd." MacCauley's head wagged back and forth in apparent shock at Peck's loss of control of his shooter.

Peck took a hard glance from MacCauley and floundered "Uhm, yessir... He left me thinking I was about to get some good news finally and—"

"And this is a major setback." Richard leaned forward, intent. "Are you back to square one? Still trying to decide between the Chinese and those environmental guys? The Room for Tomorrow guy, or chick?"

"To be honest," Peck tried again to stay out in front, "I'm leaning toward the environmental guy, Reyes. I've been putting together a financial picture of your losses and their investment activity."

"Look, Peck." Richard Abadon smiled through thin lips. "Let's just wait and see what shakes loose. If something becomes firm, make plans for, well. Let me know what you find. Focus on those RFT investments you mentioned." He put his index finger to his chin as an idea came to him. "Let's see if there are any hits when you cross major Democratic contest donors with their donor file. Individuals, not PACs." He pulled back his shirt sleeve to glance at his watch.

"Yes, sir." Peck sensed a shift, finality. Abadon seemed to have already mentally moved elsewhere.

MacCauley asked. "Sir? Is that all for today?"

They rose, not expecting a handshake but the usual cursory dismissal.

"Yes, but one last thing. If you and MacCauley are finally convinced that you have identified the thief. Make that two things—get the money back, and I want you two to move ahead with Weed Cutter. Put the Euro team on it if they travel. Just get this handled."

Peck just nodded slightly at the waist in acknowledgment. "Yes, sir. I'll put the word out."

Abadon made hard contact with MacCauley and said. "Have a good day, Mac." He didn't even make eye contact with Peck before turning to his desk.

Peck's gut boiled. *The man is again asking him to commit murder of highly public figures but couldn't even look me in the eye.* Abadon was by that time totally engrossed in thought as the flat-screen mounted flush with his desktop began scrolling stock prices. Peck turned and followed MacCauley out to the anteroom, pissed.

~ ~ ~

Three days had passed since Peck's last office visit and there had been no further word of the missing "field operative" as MacCauley had euphemized Tom Corliss. Andrew Abadon paced angrily behind his desk. He held a bank transfer statement in hand, one that left traces he'd never want subpoenaed. The elder Abadon rarely lost his temper, but the growing mystery behind their disappearing funds was frustrating. If the thieves had been content with the occasional skimming of thousands, the additional expenses of tracking down the accounting errors might not even be worth the effort. But millions?

Margaret's almost always soothing voice piped in on the desk phone. "Excuse me, sir — your brother has closed out that trip to Houston; he's on his way over." As he was answering a perfunctory thank you, both doors swung open and the younger brother, Richard Abadon, stormed in.

Richard plopped into an oversized garnet leather office chair. Making eye contact with the now stationary Andrew, he was finishing a conversation via Bluetooth earpiece and holding up a single index finger for Andrew to wait.

"Yes, Mac, I get it. You don't know where he is. We don't really care so much where he is; he's in your wheelhouse. What I'm most interested in is whether he has located any part of the misappropriated funds and has decided to change horses midrace." There was an extended pause as Richard listened to his earpiece.

"That is entirely in your wheelhouse, MacCauley. If you can find the bastard, make sure he hasn't found and absconded with any of those funds. Am I explicitly clear?" A pause. "Yes, bank records, travel records, whatever you think you might need to investigate." Another pause. "So investigate. I'm not really interested in continuing this conversation until you can tell me something that won't burn through my ACE inhibitors." He puffed in exasperation. "Look, Mac. Get Peck out of his hole and bring him back in as early as possible." He tapped the earpiece, disconnecting.

Andrew, in obvious restrained anger, demanded, "What is the total damage, Richard? How much have you identified?"

Richard, unphased by his older brother's mood, pursed his lips and sniffed, snorking up an imminent nasal drip. His own anger had loosened congestion. His proclivity to a runny nose, a result of a minor cocaine habit, was expressing itself. "If my last reports are worth anything, 243 million dollars." He huffed in sharply, in advance of a sneeze, and pulled a handkerchief from a pocket just in time. He honked loudly into the cloth, then attempted to clear both nostrils, a gesture he could do nowhere else but in the privacy of their own two offices.

Andrew's mood softened slightly. "Brother, I was just thinking that a few hundred million dollars is nothing to sneeze at. But you have stolen my thunder." His smile quickly disappeared. "How many exactly is a few hundred million?"

Richard, still clearing his upper lip, responded slowly. "243 million and change."

"You're sure that's all?"

"Hell no, I'm not sure. I've got MacCauley's people looking into the audits. He's going to start in Grand Cayman tomorrow, then move on to Geneva." Through now wet eyes, he met his brother's disapproving gaze. "Look, I'm at least as pissed as you. Peck came with high recommendations from DOD. Served with honor on several top clearance missions before his discharge. Unfortunate affair. Or maybe not, thinking about it now. And his service as a mercenary in Mosul was said to be worthy of military honors if he was in service. On the other hand, he's his own worst enemy — bad temper and a vengeful streak."

He sniffed into the back of his hand. "Peck just said his guy Corliss is missing, and he doesn't know if Corliss had back channeled any his own leads on the vulnerable accounts. As for his whereabouts, his first payment was claimed within hours; the second one has never been accessed." He refolded the fouled handkerchief to get a clean outer surface. "You saw that message, right?" He didn't pause and continued before Andrew nodded agreement. "Peck's got people searching marinas down the Mexican coast."

"Marinas?"

"He says the man's boat is missing. Marinas are pretty damn casual about boats coming and going as long as they get slip rent."

"Richard?" Andrew had cooled and had moved into calmer waters; he stepped around his chair, turned it to face the desk, and sat. "How much trust do you have in this Peck?"

"Less than I did a few days ago. But listen, he's the one that found out about the losses in the first place. He has more than one lead that he's working, and he's got to run them to ground."

"For instance?"

"Chinese, a local cybertech, and, well, Imbler."

"Our IT?"

"Yeah, he has access. He's one of the very few that have access to both banks. Short list of personnel that can access both Cayman and Swiss banks. It's set up that way —"

"Hell, Rick, I know. I set it up that way."

"OK. Well, Imbler was divorced eight months ago. Now he has a newer, younger edition he's seeing and has been treating her to weekend trips to the Rockies, rock climbing lessons, a cruise on his last vacation, and more recently, a trip to the Bahamas."

"Grand Cayman?"

"Hard to trace local moves, but it's the kind of stupid move a mid-life crisis will get you into." He snorked one nostril. "Profile of an idiot in love."

"You said there were leads, plural. Anything more?"

"Keshawn Tindall, she has access too. No profile yet."

"Hell's bells, she's what, in her sixties?" He scoffed. "Not quite your profile for an embezzler on that scale. I'd take her for the pens and pencils kind of girl, but not 243 million and change! Dammit! What about the Chinese? Or this cyber guy?" He closed his eyes, dug into them with his forefingers. Looking up with now reddened eyes, he said. "Dig deeper."

"MacCauley has investigators on Tindall nonetheless."

"Anyone else?"

"Tindall's second in command, Pradeep Rashawaran, is on the list; he's got darknet roots in Mumbai. Then there's Room for

Tomorrow. Their number two is the cybertech Peck and MacCauley are profiling."

"I read their profile in *Barron's* weekly, they had write-ups in the *Journal* too, and hell, even *People* magazine. So what?"

Now Richard scoffed. "That was more of a puff piece on the romance between the two directors. It's kind of a September-May thing." He got a blank stare back from his brother. "She's about ten years older than he is. Kind of bucks the trend."

"Maybe she has certain skills. Who cares, what about them?"

"Drew, MacCauley says their operation has grown incredibly over the last year and a half. They optioned their headquarters building after just a few months. It's hard to raise that kind of capital by making phone calls and stuffing mail. I see them as a threat at the highest level. A threat to our energy portfolio."

Andrew shrugged, "I don't know. There're a million bleeding hearts out there willing to give you ten dollars. You send out the right mail, you have ten million dollars. That's not chump change even in LA."

"MacCauley said there are bumps in their operations which follow both of the missing transactions that match our siphoned funds by a few weeks each. One of 'em, payoff on their building, came ten days after one of our hits. They closed on some farmland up the valley east of Bakersfield three weeks after another of our hits. Could be coincidental, but — ."

Andrew's eyes narrowed as he considered the implications of corporate espionage and treachery by do-gooding hippie wannabes.

"There's something else. That guy? Reyes? MacCauley said a Carlos Reyes used to work for one of our IT subs a few years back." Richard tossed the possibility out even though the chance of coincidence was high. "There are only so many hacker's hackers out there, so maybe there's something to it. How many Carl Reyes can there be?"

At that, Andrew Abadon leaned back into the plush leather of his chair and crossed his arms. "Tell Peck to work on that. Let MacCauley continue to search for more graft, in the islands and Geneva." His chin dimpled in determined fury. "I want the son of a bitch that's stealing my money. Getting the money back will be great, but I want that SOB found and fucked up. I want to nail whoever did this to a dark, damp floor in a forgettable federal prison for the rest of his natural days."

Richard relaxed into the seat, smiling. "Andrew, you don't need to know details at this time." The eighth floor of Abadon Industries was again reasonably content for one more night.

22

Parker and Carl left the Charleston, West Virginia federal courthouse wondering if the trip had been a good use of their time. Munson Mining's CEO had finally been indicted for egregious abuse of mine safety standards which resulted in twenty-nine counts of industrial homicide. The indictment had come more than four years after the incident, a testament to big money's interference with investigation and reporting. Although, some environmentalist reporting said that was amazingly swift given that other civil cases were still unresolved. After a year and all the money spent on the trial, the reduced charges and minimal six-month sentence were offensive. Coal was one of the worst polluters on the list of energy sources, second only to unregulated burning of municipal trash in Third World economies. To the energized board of Room for Tomorrow, a consensus rose for bringing twenty-nine civil counts of negligent homicide against Munson Mining's entire board. They hoped to keep Munson Mining from efficient operations for some time.

"Think we'll accomplish anything?" Parker was slowly sucking on her lower lip, a sure sign of indecision. A habit that Carl had learned to recognize months ago.

"Yeah, probably. Even if we don't put the guy away. We'll keep him from his daily routine, keep his profile high, and maybe force him to at least pay his damned fines and reparations. Realistically, if we can settle, I'd be happy if they would just agree to pay the reparations without taking each case individually to appeals court."

Parker said, "Bigelow, one of the new VPs for strategic planning, proposed suing the EPA for failure to enforce, forcing them to place liens on Munson Mining's holdings. That might be taken up when we return to Anaheim." As they got to the bottom of the courthouse steps, she stopped. She'd flipped back on the courtroom they'd just left. "But if we can put that son of a bitch away, his replacement on the board, Carraway, seems a lot more likely to talk to Tesla about the new plant. I just don't understand why an alternative use for coal, molecular-layered carbon structures, is a threat to Munson Mining. It's a final use for coal that doesn't pollute and will actually help with the energy solution. Win-win. That ass Munson can't see the nose on his face."

"It's the immediate market, hon. They can't sell huge quantities of coal for future batteries now. If you lay off workforce, in five years you'll have that workforce either retrained as something else, moved to find other work, or accustomed to taking federal money for no effort. They won't go back down the hole."

Parker said something under her breath. Carl figured it was profane exasperation. He looked at the surrounding city, re-calibrating his biological clock. "Let's go get a drink. It's cocktail time somewhere." They settled on a high-top table for two at a downtown smoke-scented lounge nestled under a parking garage. They ordered coffee instead. He took a sip of his brew after blowing the steaming edge of the cup. "When do you think we can get visas for China?"

"Don't know; soon, I hope. It's not like we don't have money to spread around." She sipped hers and resumed sucking her lower lip. She said, emerging from her funk, "P'eng probably sees us as political troublemakers. We come in, make noises about air pollution, and get some press. It's impossible to deny that there is an air pollution problem."

He looked across the street at the facing building, a modern three-story with design elements imitating a keystone arch that looked entirely out of place on a stylized lintel built of Styrofoam and stucco. A shapely young woman in professional office beige walked past, and he tracked her progress. She disappeared down the street, and his wandering mind settled on a productive thought. "What if we approached some of the companies with scrubbing solutions for those hundreds of smokestacks? Surely there'd be a net gain."

"If there was a solution that wouldn't cost money, it might happen. You have to remember, Carl, it's a state-run commercial system. A corrupt state-run commercial system. We're not going to make waves over there with cheap; we'd have to make waves with free."

"Do you think they'd listen to free?"

"Possibly, we have anyone on staff who speaks Pekingese Mandarin?"

"What about Chu?"

She laughed. "Cha Mon. Maybe Jamaican Mandarin."

He didn't laugh. "It's worth a shot. He's Mandarin Chinese. We can recruit back in LA; there's a lot of multilingual talent there. I have images in my head of millions of Chinese walking around with useless cotton masks and dying at thirty-eight from black lung. There's more black-lung in Beijing than anywhere else."

"We can worry about Beijing later, sweetie. I want to think about Paris."

He raised his coffee cup in salute. "Hell yeah, you, me, and the G8." He flashed his boyish grin. "Business and pleasure?"

Parker wasn't known for being good at impersonations, but Sarah Palin was easy. "You betcha!"

~ ~ ~

"Carl? Did you pay the bellman?" She leaned over the elegant marble sink, spit, and rinsed. Staring at her image in the mirror, she hated the "wisdom" lines growing from the inside corners of her eyes. Especially when she traveled, it seemed that her cheeks were just a little bit puffier in the morning. She smiled at the little fleur de lis on the hand towels and understated luxuries; embroidered into the edges of the St. James Albany Hotel's linens. What should she expect for a room overlooking the Tuileries? Looking out their third-floor window, she could find the Louvre to the left and to the far right, beyond the garden walls, the Seine, and the Tower.

She drew her cheeks in, accentuating the high chiseled cheek bones that were just a tad softer now. *Damn, have I gained weight?* She hated scales, and she really hated scales when she traveled. Her trusty balance beam scale at home never lied. She didn't always like the report, but you could bank on it. *Bank!*

"Carl? What was the last donation from the Abadons?"

"What?" A silent pause followed, long enough for her to be aware of street traffic two floors down. "Come out of there so we can talk without shouting through the door."

Parker smiled and put on 'the face,' turned, and went into the mahogany-trimmed hotel suite's bedroom. "I asked how much you transferred from the Abadons. I've been wondering about how smart it is to keep hitting them below the knees." She looked at Carl's tanned legs contrasting with white terry cloth and appreciated the runner's calves, one crossed on the other and resting on the low table. "If the 'big money' behind Tom Corliss was the Abadons, money doesn't get much bigger than that."

"There's the Rothschilds, Carnegie, Ford?"

"I'm not sure that old money is as dangerous or as jealous as new money."

"Dangerous is dangerous—somebody paid for a contract hit on us. The fucking Russian mafia or the Chinese syndicates? Christ!"

Carl still had his plush hotel robe on. He held the dainty gold-rimmed porcelain coffee cup from room service between thumb and forefinger with his pinky extended. "You know, at a Residence Inn you get an honest-to-god coffee mug. I'll have to fill this thing three times to get my morning shot of caffeine."

"Yeah, and at a Holiday Inn you get Le Styrofoam. Suck it up."

She watched, growing impatient as he ran his eyes down her body. "Stop it, I'm not meat."

"Oh, don't I know it. You, madam, are one hot article, certainly not meat."

She flushed, reddening in front of her ears in spite of herself. He was apparently a lot less critical of her morning self than she was. "Sorry if last night wasn't enough to cool your ardor, suh." She took a pose, hip out. "I do try to please mah menfolk."

"Au contraire, mademoiselle. If anything, you stoked the fires within." His accent, by comparison, was painfully faux Parisienne.

"You are hopeless." She took a cup of from the serving tray, filled it, and turned to face him.

"Why do you say hopeless—a man *has* to maintain hope." One side of his smile raised in a devilish smirk. She looked for signs of the previous night's conversation on the future of their 'relationship,' as she'd called it.

"Don't go there again, Carl. I had to divorce the second one, and the first one died on me. I'm not marrying another." She set her face in a smile but not one with implied invitation. "I love you; you know that. I'm pretty sure you're crazy about me too. Think about it. *If* anybody knew about our freelancing, we'd both be world-class wanted felons on most of the planet's continents. There's no telling what next week, month, or year holds. Ask me again in five years if neither of us is in jail." She started toward him but paused a step away and sat on the damask tapestry upholstered settee across from his.

He set his own face in a smile too. His seemed to be one of resignation, acceptance.

She thought he'd delivered some heavy arguments for marriage last night. He did love her. Her audacious determination. Force of nature or force for nature, either way. He'd take her either way.

He answered the earlier question. "One hundred, twenty-one point seven million dollars." He traced the numbers in the air in reverse as he slowly sounded out the numbers. "It duplicates a transfer of funds from one of the 'excess profits' accounts in Switzerland to another in the Caymans. It will look like an error next quarter when they're balancing the books."

She nodded appreciatively and blew a silent whistle. "You'd think an error like that would make someone notice. I think we should be very careful." Something in his face had just changed, hardened.

"Very true. I agree. There are a lot of rich, unscrupulous bastards in the world. Take that fat schmuck back in West Virginia." Carl was still pissed at the light sentence, and the wink and nod judge. "The grand jury failed to make him a poster boy for world class bastard. Most of the world assumes he's a lying cheating bastard. And his good old boys in the Old Dominion's fraternity clubs are grinnin' and spittin', bragging on him."

"You want to go after him?"

His face took on the brow-furrowed air of a conspirator. "Yeah, and his ilk. We'll call it the greedy bastards club. A secret fraternity whose only invitation will be the noticeable transfer of wealth, a great deal of wealth, to the Caymans or their friends in

Geneva." Carl got up to refill his undersized coffee cup. "I think you are right to want to keep clear of Abadon Industries. This last little bookkeeping error will be the last, and we can get a lot of good things done with it. We're going to have a hefty engineering bill to pay in Houston."

Her tiny cup empty already, Parker stood and followed him to the breakfast tray. She put her hands on his strong shoulders, gently so he wouldn't spill his coffee. "Pour me another one, dear." She passed her cup around him on the table. While he was pouring, she nibbled at the base of his neck while she undid the tie on her own robe and let it fall open. She pressed her soft contours into the firm muscles of his back, grabbed the edges of her robe, and pulled them forward so he was included in the wrap. She cooed into his ear, "Carl, why don't we reconsider that breakfast menu."

"What about the Eggs Benedict, honeyed French toast... we've got two juices and..."

"Carl, shut up and turn around."

He did so and felt her fingers untying his robe's belt. As she slid down in front of him, her fingers trailed across his chest, then down and traced the angry red welt where Corliss' bullet had creased his beautiful skin.

His gaze wandered across the visible wedge of the eternal city spread out before him. His second-floor view took in the steeply slanted rooftops of the Paris skyline across the Seine. Even at the edge of his view, the Tour de Eiffel demanded to be the focal point. This would be a moment to remember forever.

Her fingers had just crossed the reddened thickening that marked the bullet hole in his side when an explosion in the street below lifted the wrought iron balcony railing up and into the suite's window glass. The first blast wave almost missed them, blowing vertically past their wall due to the angle but sending glass shards through the room. They were thrown to the floor into a pile of upturned furniture and the breakfast cart. The second blast from the truck bomb across the street reflected on the heavy Tuileries perimeter wall and sent a searing blast of heat into the room.

23

Sirens in that particular European hi-lo wee-waah cry responded to the two blasts immediately. The City of Lights had become accustomed to tragedies. For Parker and Carl, it would take a while longer. The wail of sirens barely made it through the temporary high-pitched whine, the only thing heard by their stunned cochlea. They lay tangled half-in and half-out of their robes among the ruins of breakfast and the relocated furniture in the suite. Bits of fabric from shredded curtains floated down, motely remnants of luxury.

Carl spoke first, inelegantly responding to the unreality of the explosion. "Dammit!" Then through a groan, "Oh NO!" He could feel her weight on his legs, her heart pounding against his thigh. "Are you all right?" He felt around for immediate injury, aware that shock would mask the initial sensations. He felt at a bump on his forehead and picked out a bloody shard of glass. Her hair covered her face. Her legs curled unnaturally from her robe. His first stroke at the back of her head found a piece of glass shrapnel. His stomach curdled in fear. Getting into a sitting position, he carefully pulled at the piece of ancient window glass. It came free, its tip coated in her blood.

He asked again, "Park? Can you hear me? Are you all right?" The question felt absolutely stupid. *Of course, she's not all right.* "Park?"

She groaned softly and then, "I don't know, I think so." She sat up, feeling for anything obvious. "Christ!" Finding glass in her forearm, she began carefully picking at sparkling shards. "Fucking Jihadists!" She looked up, met Carl's stare. "Oh hell, Carl."

"What?"

"Don't touch your face!"

~ ~ ~

Sunlight might have made the room a cheery place to recover from the shock and injuries of the bombing. Unfortunately, two days of unrelenting overcast and occasional drizzle painted permanent shadows in every corner of the room. Bouquets sent from Anaheim and get well notes from some of Room for Tomorrow's most reliable benefactors helped perk up the stale atmosphere. Parker flicked through the channel listings again. There were two BBC channels somewhere in this damn lineup! Damn meds! Channel seventeen? No, twenty-seven? What? Wait. She hit the back button, and seventeen had just gone to coverage of the bombing on the Rue de Rivoli.

Blast damage devastated the lower floors on two blocks. Destruction of the first floor was complete with *LeMonde* reporting that glass from the hotel's chandeliers had embedded in the plastered stone walls supporting the central corridor of the hotel. Postcard street vendors, venerable staff, and world class guests all equalized now in death. A section of the stone wall blocking street noise from the Jardin des Tuileries had been blown into the gardens. The normally crisp Paris Police uniforms were now replaced by riot gear and automatic weapons slung from shoulders. The metro station was closed off from the street. Her high school French and a college refresher gave her only a rudimentary understanding of newscaster's speed French, and she hated missing most of it.

The videos were gruesome. Bodies under bloodstained sheets, wounded being loaded onto ambulances. The scenes reminded her of too many around the world where economic frustration fueled religious fervor to convince the gullible that God would smile on them and their families, if only they would blow up fifty of the infidels.

The actual body count in Paris had been higher, last reports indicated sixty-seven killed outright, forty-nine more hospitalized with thirteen of those in critical condition. Tour buses waiting at the corner had been about to leave for Versailles and the carnage aboard one of them was horrible. Over thirteen dead, twenty-five more sent to hospitals from the buses alone.

Listening carefully, she thought she heard the announcer say if the morning Metro had not been late, the number of casualties on the street would have been much higher. She pictured the view from their balcony. Yes, the Metro stop was just across the street.

Beyond its stairwell, a portion of the garden wall was a pile of rubble. The wall's solid masonry must have been part of the reason for the reflected blast that sprayed glass into their room. She hadn't seen anything. The explosion had blown her down over Carl and stunned them both. She'd thought she was going to be deaf until his voice had brought her back. Even now though, she could hear a constant ringing in her ears. The voice on the set came back into her consciousness; she turned the volume down to just above a whisper. The scene had moved to a desert climate, Al Jazeera's logo was in the upper corner, and an Arab talking head was now reporting. That translation into French was hard to follow, but the text scroll at the bottom gave her a little more to understand. Not taking direct credit, ISIS was giving credit for the attack to the AZM.

She repeated her earlier summary judgment. "Fucking Jihadists." She clicked off the wall-mounted set, slid off the high mattress, and taking the IV tree with her, walked over to look outside. Thinking back to the telecast and Grieg's vid of the Last Day, she thought to herself, *No wonder everyone wanted to make sure you were blown off the map for good.* She looked up at bands of light in the overcast. Maybe the cloud cover was going to break soon. The modern office buildings surrounding the American Hospital belied her location. It could have been almost any modern city. Well, maybe she mused, with a few more Citroens on the street than others. She jumped at the sound of tapping at the door. Carl stood in its frame, hospital-robed, smiling, and grasping his rolling IV stand. His face was a patchwork of white butterfly bandages pulling together the multiple cuts from flying glass.

"Hey, gorgeous."

"Hey, handsome." She tried to smile, but her face hurt from hitting the breakfast cart, then the floor; and her own cuts from glass shards peppered every movement of her neck and shoulders. Her emotions were boiling; she had wanted to say something cool, diffident, or enraged and engaged. Instead, all she wanted to do was hug and be hugged. What a comic scene! The two of them rolling their saline drip paraphernalia to meet at the end of the bed and embrace, being careful not to entangle each other's plastic tubing. As she laid her head against his chest, he was humming. *What's that tune? Oh, the Dido love ballad, White Flag.* The melody was slow and haunting, he could be such a mush at times. His hand slowly slid down her back, a little too far. "You're terrible!"

"That may be, but why do you think so?"

"I can barely hobble, especially with this contraption wired to me, and you come in buzzing with music and, and, carnal intent."

"Was I?"

"You know you were."

"Sorry, and I mean it." He guided her to the bed and tried to make her sit. He said, "I've been doing a little research. The police are happy to have a public enemy to blame, one that is easy to blame, but investigators are wondering about the claim that it was the AZMs."

Restless from bedrest, she went to the window instead and looked out at a dismal afternoon. "But ISIS just said the Anti-Zion Militia claimed responsibility. I just saw the clip on TV!"

"But the defense ministry is wondering why the explosive residue matches samples manufactured in the US."

"Couldn't they have gotten the explosives from unexploded ordnance? Lord knows we've been hauling boat loads of munitions to the Middle East for decades."

"Not American military. Its American-made but from a batch made for mining. They're working on an exact batch ID, but it seems it may have been part of a shipment sent to Kentucky and West Virginia — to a major coal mining firm that you may be familiar with."

She stared, slack jawed. "It couldn't be. No, that can't be." She turned away from the window. A break in the clouds had sent a beam of light in from a glass-paneled building across the boulevard, splashing her in a golden wash.

He continued to detail what might be bad news. "It's close; enough of a match that the General Directorate has asked NSA for assistance in getting a warrant."

"They can tell that?" She needed her rolling IV rack for support, her knees suddenly weaker.

"Chemical tracers. Who do we know that makes a lot of chemicals?"

"Don't say it, Carl. I won't believe that even they would risk blowing up a hundred innocent tourists and maybe destroy a Paris landmark building just to kill us." She shut her eyes tight, forcing the tears to stop. "We're just a couple of do-gooding enviro-whack jobs to them." She entertained some disturbing thoughts about the traceability of the claymores they had dropped on the North Florida Gas Pipeline.

He bent over, kissing the tear tracing down her left cheek. The pipeline memories melted away, and she lifted her mouth to meet his. The ardent sincerity in their life-affirming kiss gave her second thoughts about her refusal to discuss marriage.

He pulled back, meeting her moist eyes. "We're more than do-gooding whack jobs. And maybe they would try harder if they're getting any better at accounting."

His fingers began to tap out the frenetic beat of a Jan Hammer soundtrack. She wanted to ask something about actions and reactions, or karma, or consequences. Instead, she found her chest heaving and shuddering as uncontrolled sobbing took over. Carl sat on the narrow edge of the bed and embraced her in all the hug he could muster in the limited slack from his IV tree's tubing.

~ ~ ~

Neither of them still needed bandages when they made their welcomed return to the bustling RFT headquarters in Anaheim, but both of them needed foundation makeup to calm the troops. Parker had spent an hour finalizing makeup to conceal the remnants of a black eye. Carl's forehead still showed red pocks on his left side from glass splinters. As they entered the RTF cafeteria, the assembled workforce of 220 stood applauding, cheering. News of their hospitalization after the terrorist bombing had sent understandable shock waves through their building and the worldwide environmental movement. Rumors of a targeted attack were squelched by RFT's PR department at Parker's specific request.

One of the administrative assistants entered, rolling a mail cart with three stuffed plastic mail crates. Terry Bouschet, VP for Marketing and Outreach spoke up. "Parker! Carl!" He opened his arms in grand eloquent welcome from across the table. "You have mail! And lots of it. You'll probably want to run a filter on your email inbox 'cause we had to disable your personal emails on the server. Seems a whole lot of people are glad to hear you were safe. Absolute hit on the net. Your injuries were the fourth-most trending topic on Twitter; the bombing itself was first. And Facebook, well…" He went on ebulliently, as only a trained salesman can do, long enough for everyone else to file by and hug or shake hands. When Joan Brayton made her close-in approach for a welcome, she hugged them both. She whispered in Parker's ear. "We need to talk after we're done in the boardroom."

The tinkling sounds of the recently completed boardroom's water wall feature helped to calm the mood, but none of the board's members had taken a seat. Parker raised her arms, revealing a few remaining bandages. "Please, please! Seats everyone. We have work to do and some announcements. Sorry I didn't get an agenda to you. The plane's Fli-Fi was down. Thank you all, but it appears we need to get back to work." She looked up at Carl, a slight rim of moisture in her eyes.

Carl looked over the assembled board members. Parker sat opposite him at the conference table as was their habit. He began. "Friends, other than our little delay in getting out of Paris," an array smiles and nods responded, "we had a successful trip to Europe. I think you are all aware of the wind farm in the Thames estuary, the largest array in Europe to date." Most of the heads nodded; some did not. "OK, their 175 turbines together can produce 630 MW of power. That's significant — that turbine array can power two-thirds of Kent on a good day." He waved a hand toward the bronze-tinted window, "And unlike our Mr. Sun, the wind blows at night."

"But that's a done deal. They gave us a nice tour, and the bangers and mash, and the wee bit of local brew were great." He paused for a few chuckles to settle. "We also spent time a little further north, meeting with GE's Haliade 150 development team. At about 240-feet in length, its blades hold the current record for size. Their designers were very happy to hear from us. They have been seeking funding for over two years for a major turbine farm in the North Sea. Euro troubles it seems, with Brexit complications.

We have a deal, but the deal is a great one for the US. Commensurate with our funding the North Sea Project at thirty percent, their patent is to be registered in the US, with new manufacturing plants set up in Rust Belt communities." A murmur traveled around the room. "Yes, I know, a lot of you would like to see them built here in sunny California. I'm thinking of Illinois or Indiana, even Ohio. The blades are usually shipped in two pieces on a double flat car. Last mile delivery on special semi-rigs. We won't be able to ship if there are underpass bridges involved."

"What about an installation here?" An unidentified voice broke in.

"We have winds, but we'd have to set them on the ridge lines. Those ridge lines have views and are very expensive. And in spite of the liberal attitudes usually prevalent around here, there'd be hell to pay messing with the view." He shrugged in apology. He caught a hand being raised tentatively. "Sandy, question?"

Sandra Templeton from Systems Analysis asked, "Where do you want to build the farm?"

"Ah, glad you asked. In the very heart of fracking country. The Dakotas. I want to make a point about the efficacy of wind and where better to make that point? With a wind farm the size of the Thames estuary and turbines on the order of the Haliade, we could power both Lincoln and Omaha and export power to the twin cities."

For about twenty more minutes, he sold the board, which was an unnecessary exercise since he had the numbers available and there was a line on preliminary funding. Even without an Energy Department grant, he knew he had alternate sources of money if his access to PeMex accounts was still undiscovered.

The other major board topic concerned Tesla Motor's reports of a breakthrough in the development of layered-carbon batteries — lead-free batteries the size of a suitcase that could store the energy a solar powered three-bedroom house could gather in an eight-hour day. At present they had reduced the demand for hard-to-get Lithium by 25% in the doped carbon layers. More work was needed to take the new layered-carbon batteries to market, but Tesla's first installation in a car increased its driving range per charge to over 420 miles.

Parker listened to the last discussion with drooping eyelids. The biggest problem with becoming acclimated to the European sun cycle was re-acclimating to Pacific Time. She failed at disguising a yawn and felt a growing ache behind her eyes as exhaustion began to take over. She was grateful when Carl brought the meeting to a close. She watched the members of their board, a very solid and hardworking group she thought, file out of the room. She was thinking back to when she had chased Carl down on the Coastal Highway. Her thoughts drifted to their first night, her first—no, third— visit to the old farmhouse, and the yester-today.

"Excuse me." Joan Brayton, second in overall security but head of IT security, was standing, aware of the red-rimmed exhaustion her boss was feeling. "Ms. P? I really hate to bring this up. But I feel it's urgent, and I need to advise you and Mr. Reyes before close of business. I was afraid you might head home early after your long trip."

Parker dabbed at an overworked tear duct that was trying to betray her state and pointed to the adjacent chair. "Have a seat. I can take a few more minutes." A yawn took over, and when she regained control of her face, said. "What's up?" Brayton looked across toward Carl who was being bombarded with questions about the blast. "Mr. Reyes? Can you take a minute?"

When he'd taken a seat, Brayton started in. "I'm afraid we've had a breach in the firewall."

24

"Oh! It wasn't a serious breach." Brayton reacted to the alarm registering on their faces. "But yes, we did have a breach and a denial-of-service attack on our telnet. Call center operations went down for a few minutes. At first we thought we were only taking a denial-of-service attack."

"What did they get?" Carl's knowing glance told Joan that his mind had gone immediately to server architecture.

Parker was instantly wide awake.

Brayton splayed her fingers wide, taking mental stock of the totality of the attacks. "We think only some mail lists, but in the barrage, someone was knocking on the door to the firewall...like, knocking real hard. I had to personally swat at the incoming attacks."

Carl asked, "You're sure they didn't get into our financials? The second set?"

"Yes, sir." She swallowed hard from nervous jitters and grimaced at a tidbit of bile and hoping to hell she was right. "The financial side is airtight. It has no real connection to our other systems, not one wire, Wi-Fi link, or Bluetooth device. No way they could have made that jump."

Parker asked. "Joanie, you think the barrage attack was cover for the firewall attack. Hoping we wouldn't notice?"

"Yes, ma'am." Joan Brayton was young, intelligent, and not at all a caricature nerd. Her father had encouraged her math skills while her mother introduced her to modeling and bit-part acting. In her high school sophomore summer, she and a girlfriend discovered a homeless teen's body sprawled like a discarded doll, partially concealed by a shrub. She Googled stats on the girls who came to LA looking for a big break. The scale of human misery she found in trafficking, abuse, death by overdose, and simple despair convinced her to stop modeling and take coding at the High School for Arts and Sciences.

Within a year, she had turned the high school's website into an enterprise site that took orders for yearbooks, band instruments, and graduation gowns; booked proms and dances; and even sold bumper stickers and logo clothing. Her slender five-five frame would never be imposing, and her looks would always be disarming, but, as Carl had learned, she should never be underestimated. He had insisted that she take a four-month cram course at his favorite upstate New York hacker school, not that she was expected to learn how to hack with the best, but on the bet that she would recognize attacks and stop them.

Joan Brayton took a sip from a bottle of water to clear her throat and regain her chain of thought. "You guys are celebrities now; you are world figures. Have you considered the possibilities that the Paris attack was more than a wrong place, wrong time deal?"

Carl and Parker looked at each other. They had been discussing it after a failed nap on the non-stop hop from LaGuardia to John Wayne. Despite the news of the chemical tracer study, it had seemed too far-fetched. Carl answered her. "Joanie, we gave it the three minutes it deserves on the flight back. But seriously?"

The young tech wiz persisted. "I would love for the Paris bombing to have been happenstance, a lucky near miss." She looked down at her fingers, stalling before broaching her biggest concern. "And maybe there is no connection to that and the Internet attack. But there's the coincidence of timing...with Carl not in house." Brayton smiled, changing gears, and delivered her good news as a modifier. "On the bright side, I have three teams assigned to throw water on the flames at the hostile opposition's blog boards. They only use laptops and move around coffee shops in five markets."

"And?" Parker drew her out. The pregnant pauses were alarming. "What's your gut?"

"A lot of people would like to see us go down in flames. They think we are in league with the devil, anti-church, anti-God. Some of the crazies have been asking for well-aimed bullets as a solution."

Carl objected. "But those are blogosphere blowhards. You don't take them seriously?"

She dead eyed her boss. "Sir, you as well as anyone know that some of those blowhards are pros with a purpose." She blew out a breath through puckered lips. "I'm just saying that it's the kind of thing we need to keep monitoring. And…I want to scan your cars, houses, and phones." She waited a beat, then held out her hand. "Sooner than later."

Parker's phone was face down on the table. She pushed it toward Brayton. Carl reached into his coat pocket for his.

"Joanie, can I get it back in the morning?" Parker asked. "I feel naked without it, vulnerable."

"Sooner. If you'll at least take a potty-break, I can get them back to you in fifteen minutes." They looked at her, questioning. "They're iPhones. I need a pick to get in; otherwise I'd only need a few minutes each."

Carl said, "Keep 'em. I don't want to talk to anyone else tonight; we can take a cab to the Montage. I really just need a place to shut my eyes. Not that good at sleeping while moving."

"Thank you. I'll come out to the houses and scan all of your vehicles too." Brayton scooped up the two phones and slipped them into her zippered organizer. "Sorry for the inconvenience."

"No, Joanie. Thanks for taking the initiative." Parker was proud; proud of the young woman's dedication and glad she was working for RFT and not the opposition.

They had started to leave and were nearing the elevator when Parker stopped. "Joanie?" Brayton's attention perked. Few people used her given name; she was Brayton to almost everyone in the building. "How long till you can get the houses and cars swept? That's on the order of a couple of days, right?"

Brayton looked up at a ceiling light as if calculating equipment purchases and personnel vs. contract. "Tomorrow probably, two days tops."

Parker stood, head tilted, thinking. "Joanie? When you're through with us, I want you to hire and organize a team that scans the offices, and the homes, and vehicles of anyone here above trainee on a frequency of at least twice a week. When that's up and running, we'll expand to the trainees." She turned to go but stopped and looked back. "And, come see me first thing; we'll talk about what you'd think is appropriate for a raise."

Joan's eyes widened, followed by a toothy grin. "Yes, ma'am."

Parker turned her head up to meet Carl's eyes. "Three days at the Montage will be just fine, don't you think?"

"We'll see, and I think back in Paris, before we were *so* rudely interrupted, you were about to make me an offer I shouldn't refuse." His smile broadened, then winced as the several lines of scar tissue on his face pulled on healthy skin.

~ ~ ~

One day at the luxurious Montage Laguna Beach Resort was enough. Both wanted to get home to normalcy, to reclaim stability and sanity. The last luxury hotel they'd shared was a smoldering horror show. Back at her La Jolla cliffside home, they'd settled back into comfort and finally boredom while their cars were being scanned. Carl argued that room service at the Montage delivered terrific lobster and sirloin and offered to go back, but late morning found that one of the new executive level fleet Teslas had been delivered by the ever-efficient Megumi Sakoda. The dark metallic blue machine, still shy of its RTF green repaint, took them down the coast for real Mexican food in Tijuana less than half a mile south of the border crossing. A simple text exchange with Grieg closed with a request for them to drop in if they could. After siesta time, lunch, and a beer they decided to just go let the wheels spin. and near dusk they ended up in the neighborhood of the little farmhouse south of Neenach.

They pulled into the backyard beside Corliss's abandoned vehicle. To prevent it from attracting locals, it had been stripped of radio and tires, and a few windows had been smashed. It needed to look abandoned. Parker blinked at the sight of the car on blocks. The memories flooded back, pushing the omnipresent Paris reruns to a back channel. "Do you think Corliss lasted very long?" As time separated the anger of that night from the present, Parker had begun to think his sentence to a toxic death in the future had been on the cruel side.

Carl shook his head, no. "Look, we gave him hope, and a chance. Right?"

"Hope yes, but not much of a chance." She remembered the fight scene in too much detail. "How's that stomach wound?"

"Only hurts when you're on top."

She reached across the console to pat his thigh. "I'll make it all better when we get back to the house."

Entering the farmhouse, the set of worn lacquered steps, the hallway, and the closed door at the end were by now very familiar. But as they approached the top of the stair, soft crying sounds came from a closed bedroom on the left. The door had normally been slightly ajar. They'd explored the entire house on previous visits with most rooms being entirely empty unless a few high school kids visited and left their detritus. Carl carefully pushed the door open, revealing nothing. As he stepped into the room, soft moaning began further down the hall.

Parker, still in the hallway, pushed another door completely open without entering. The moaning continued but from no apparent source. Puzzled, but aware that the house had secrets, they continued down the hall to the last door at the end. But now the little-used last door on the right, a closet in former times, held the curled body of a young teen and a tattered backpack. Parker licked the back of her hand and placed it under the girl's nose. "She's breathing." Other than this new visitor, there seemed to be no apparent differences in the upstairs. Taking the now practiced precaution of entering one at a time, they stepped into the front bedroom and through the portal.

"Grieg? Chu?" No answer from the empty rooms. It didn't take long to determine that no one was in the time chamber.

"Well? What do you want to do?" Parker took one of the two chairs and sat at the small table. "Wait on someone to come back or keep moving?"

"Hmm, we don't know which side of the gate they're on, or when either of them is due back." Carl gestured toward the shimmering entrance. "We could always give the new visitor a ride back to town. Drop her at a runaway shelter."

Parker raised eyebrows, understanding of his past filling in any question of his concern for the young girl. But she could wait; if she'd been given the normal dose of forget-all drugs, she'd be out for hours. She nodded at the decon chamber. "You ever been curious about what's on the other side of their back door?"

"Of course, but...no, don't think we should, I'm guessing it's a sight worse than Beijing in August. That thing is more than just a shower."

She had seen video of the throngs of pedestrians in Beijing's smog and tried to compare it with the worst of a Los Angeles heat inversion nightmare. *God! Can we save us from ourselves?*

Carl continued. "They said it's to keep from contaminating the past with whatever is out there."

Parker thought about that. Then, in a return to her lawyerly habit of examining every statement at its worth, she said, "What if it's to keep from contaminating the future with *our* microbes? What if their whole line about the future being so horrible is just to keep us from sniffing it out for ourselves?"

"Wrong tangent." Carl had come to appreciate the offshoots and side roads that Parker's imagination took. It made her much more interesting. But this one didn't ring true "If the future wasn't so screwed, why would he have given us those data cubes?"

"Oh, right." She still stared at the chamber door.

"You'd really like to see, wouldn't you?"

Parker looked down at the small cabinet by the dinette/worktable. She opened the half cabinet door and saw the small drawer she knew had the knurled knobs that were supposed to be the time controls. Self-consciously looking back at the decon chamber, she reached and tapped the drawer. It silently slid open. Above a patch of fine print were five knobs. Labels were set for time intervals in decimals. Four were fixed with a screw through their flanges that kept them from turning. Only one, labeled "Hours" and showing a range of -36 to +36, appeared moveable. "So, this is why they say CalStation can only go back a day for do overs."

"Parker, what the hell are you doing?" Carl glared at her in disbelief.

She nearly jumped out of her chair when the decon chamber's light came on, accompanied by a soft audible alarm signal.

She slammed the drawer shut and fastened the hasp on the cabinet. Carl laughed. "Ha! Busted! Just for thinking about it." They watched as a masked and suited figure came through the outer lock. The air visible on the other side had been dark, almost thick with dirt. They had arrived in what passed for great weather in Southern California. A series of indicator lights lining one side of a control panel slowly sequenced from red to green. With a small gasp of negative pressures, the door popped open.

223

Bruce Ballister

25

Chu removed the filtration mask and turned to put it down. He jumped slightly, startled to see Parker and Carl sitting quietly at the tiny common area table. "Klaatu barada nikto." He waited for a response and repeated, "Klaatu barada nikto?"

"What?" The two visitors shared a silent question, then Carl slapped his knee. "Oh! From *The Day the Earth Stood Still*, Michael Rennie, uhm, '60s I think."

"Ah yeah. Damn, that one is still a classic." Parker offered.

Chu pulled a bunk out of its recess in the wall and sat. He took in the remaining scars and bandages. "You two look like you've been messing with the bees. What happened?"

The two visitors shared a glance and a grimace. Carl shrugged, said. "Long story."

"Been here long?"

"No." Parker spoke. "We just got here actually. We had taken a few days off after a tiring trip and got bored, took a drive, got Grieg's message, and here we are."

"You're always welcome, of course, but you know, your car, even in the back, could lead to more visitors. Tire tracks."

Parker and Carl exchanged glances.

Parker made a meaningful glance over her shoulder to the portal. "You may want to go check outside in the hall. I think one of them is having some trouble figuring out what just happened to them. Don't you take them back? I woke up back in my bed."

Chu exclaimed, "Buddha balls! We didn't know where she came from, no ID. Maybe a runaway. Wherever she gets to in the next six or eight hours won't matter; she won't remember anything before then, especially here?"

"If she's a runaway, won't she just stay the night?"

"Right. Bugger it all." Chu pulled at the hint of a beard on his chin. "You blokes know your century. What would work without unnecessary mayhem?"

"You sure she won't remember?" Carl asked.

"Absolutely. Well," Chu paused, tilting his head toward Parker, "with some great certainty, less than 100 percent, you are an anomaly. That cocktail? You had asked about it before. It's a little bit 22nd century pharmaceutical magic, some 20th Century Ativan, and, since our sources are limited, bit 14th century Chinese horticulture. The street name is voodoo déjà-vu. I hear it give you dreams to make the willies jump right outta you."

"Voodoo déjà-vu, quite a handle," Parker said, appreciating the street label which was new to her at least. "Who came up with it?"

"Not sure. In my time it's been around awhile. Came out of the Triad's research in Hong Kong, before the wars." He remembered what the essence of the problem was. "You said she's still outside?"

Carl answered, "Someone, a teenage girl, is semi-conscious, right outside the door. Southside bedroom." On further consideration, he said, "Don't worry, Chu. We can deal with her." He checked sidelong with Parker, "We can give her a ride down the mountain, right?"

"Sure," Parker said, then to Chu, "Has anyone else returned, like I did?"

"Only a few. A park ranger was the only uniform. We got lucky with you two. But you're not typical of your times. You are both very bright and talented."

Carl bowed slightly. "Thanks." He then thought about the insult to all the slags in his time.

"No, serious. I'm not sure if I've said it, but I don't know if I can accomplish what I want to do here without your help."

"You're trying to avoid a global climate catastrophe that leads to global thermonuclear catastrophe. That's not a hard agenda to sign up for," Parker replied.

"But," Chu hesitated. "That wasn't our original mission here. I think you know that." He saw affirmative nods. "We're scavengers in the digital dust bin. We lost so much. So many—" His voice trailed off.

"Listen, Chu," Parker said. "I couldn't be happier. In a little over a year, Carl and I have bootstrapped a major environmental movement that's ostensibly funded by tens of thousands of individual private donors. We're making some positive changes. It could be a self-sustaining legacy even if we stop ripping off the Bigs!" She stopped—a nagging question that had been worrying her bubbled up. "Is there any way for you to know, you know, by looking out your back door. Is anything we're doing making any difference at all?"

"Well, one or two things would tell you right away. You come here, and there's nothing but an old house, no time gate. Something's changed that means the 22nd didn't set up a time gate. Another outcome is you come here and other guys are here doing simple research. They wonder who the hell you are. The 23rd is alive and well, or at least not decimated. Either one of those, you know you did something. Right or wrong, good or bad, you did something."

Parker's face wrinkled in frustration. "I hope, well, I don't know what to hope. It's just… dammit! I don't want everything I know to go to hell in a bucket." She stole a quick sideward glance at Carl when she added. "If I were to raise kids, by adoption at my age, I'd want to know that their kids' and grandkids' efforts weren't going to be fried, radiated, or sandblasted to hell."

Chu gave the twin shoulder shrug they had learned was 23rd century body language for sure. Carl asked, "When the 22nd set up these stations, how many were there? I'm guessing there were more than this one."

Chu glanced down, paused, then, "I don't know, Carl. I do know there's this one. There are five of them melted to slag near major cities. There's the first one we found outside Queensland, radioactive. But it showed us this one, and there's another one in Johannesburg. When we bounce a signal to it, life signs say its operational like this one, but the city is too hot. We think there were more in the network, but we haven't found a complete map." Parker opened her mouth as if to speak. "And before you ask, yes, Johannesburg and Queensland are hot. So hot, nobody who's ever gone to check ever came back widout de Big Luke." When he only got blank stares, he added. "Leukemia."

They both grimaced at the prospect. Carl pursued his line of inquiry further. "Does this station have a dial or adjustment that takes it to other times, er, different times?" He knew Parker was not about to fess up to her earlier snooping.

"Want to do some time hopping?"

Parker looked at him, a picture of innocence.

He pursued his side question, one that was important to his fund-raising efforts. "If I went back ten months, I could make some good stock picks."

Chu countered. "Yeah, and if you went forward ten years you could see if you got killed doing this or not."

Carl coughed uncomfortably. "There's that. But it could help us make better investment decisions too."

"Your answer, I'm sorry, is no. We only understand the basic mechanics of time displacement. I'm a user, not a designer. Neither of us know the metaphysical paradigms that permit dwelling in two time streams. Those are lost. We can play with a day or two at most, and that's risky. The long-term verniers work, but they're screwed down to keep us from screwing up." Chu shrugged. "I don't know, I think the original designers could have time shifted almost anywhere they wanted; we don't know how. We don't know why this one was set for the current time displacement, but it's damned useful to us for the built-in filtration system."

Carl asked. "Then, you can only shift within a few hours? Only one vernier works?"

Chu gave him a penetrating look. *How dis boy know dat?* "You do know that your time is elapsing at the same rate as mine? This is a temporal displacement gate. There is always same displacement across de barrier; more or less. If we could monkey with those dials, it might be different. If I could have figured out how to power down the time field effect, we would have." He shrugged. "To keep complications like you and that girl in the hall off our to-do list."

Parker had to think back to the yester-today Sunday. "When you guys took me back, you put me back a whole day plus a few hours. I did that Sunday twice."

"Yeah, we can only do a little over 24 hours. It was an experiment really to see what would happen. We decided it was too metaphysically buggered to try to do more of it. I didn't want to see myself comin' and goin' and wonder who should or shouldn't be there. It could go to the wops fast."

"Damned shame." Parker had the same questions about the dial, the possibilities of going some when else. *Was that a word?* she thought. *Do time travelers go some when else?* She said, "Chu, I know your time, the 23rd, is pretty fucked up."

Chu almost laughed. "Yeah, nigh well bollixed is right."

"What you've told us tells me that humanity is pretty scared that it might not produce a viable next couple of generations." She looked back toward the time gate's portal, then back at Chu. "Would you be interested in staying here, uhm, in staying now? In the 21st?"

~ ~ ~

Chu's mouth frowned but his eyes sparkled. "Interesting question, Parker. I'll let you know what..." He lowered his head, picked at a fingernail. They watched as his chest heaved a few times, then settled. When started speaking again, his voice was uncharacteristically full of emotion. "Matter of fact, I got nothing back home. I'm an orphan, and NZ is nothing more than a first-class refugee camp. Complete with hunger, shortages, a thriving black market, crooked cops, low birth survival rates, and short life spans. Cancer eats every good thing. That's one of the things I'm trying to finish up for them, some of the cancer meds." He swallowed and continued, more in control. "Me and Grieg, we talked about it, about staying. We can and do transmit the data back to NZ. No need to go back. We could tell them we're too sick to come home. Don't send anyone."

"Isn't your tour, your assignment, almost over?" Parker felt her own chest thickening with emotion. The usually reserved Chu had shared more in thirty seconds than he had in six months.

"Yeah, well, we should have been gone months ago, uhm, but we lost one coming over the mountain, first week. Then—" He swallowed hard. "—Noel. Grieg's work took longer than expected, and our little trips to Latvia and the rest on your time took real time in both timelines." He stopped rambling. "Sorry, not the answer to your question. I, er, we've been considering it. We might not get the choice. I hope not."

"How can you mean that?" Carl asked.

Chu curled into a cross-legged half lotus, cat smooth: the shift in sincerity was stunning. He blinked, paused, said. "If you do something significant soon, or if you have set something in motion that has upstream consequences, then I might simply cease to exist if I'm on my side of the gate. If I'm on your side, it's anybody's guess. Noel said once that we might just go 'pfft' and disappear. I'm not sure that I believe that; I don't know that I believe in the construct of parallel universes or timelines."

Carl had another more pointed question in his original line of questions, and it had been side-tracked. "Chu, if we stir things up for ourselves...if we've put ourselves in very real danger in our here and now, can we get to a safe zone in your time, New Zealand 2214."

"It would be 2215 now. Remember? It's a year later there too. There's room in the boat. I'll let you know. But I'm not sure you're going to want to make that change."

Parker said, "When I make choices, I like to know what the shape of the deal is. We see you guys pass through that door, and decontamination process and — "

He grimaced, "Parker, you've never been out the back door or seen my reality."

Parker frowned, looked at the entrance to the decon chamber. "What's out there, Chu? How bad is it?"

"Hmm, I wonder if you could wear Noel's mask?"

"Really?" She looked at Carl. He shrugged a why not but his eyes betrayed worry.

Chu uncurled and stood. "Cha, sure. We tighten up the neck strap a little." He sidestepped and opened a small cabinet next to the back door, the entrance to the decontamination chamber. The mask looked like a sci-fi props designer's dream creation. The rebreather with filter did have a passing resemblance to any standard gas mask, but the goggles above featured six small armored lenses, arrayed to provide full panoramic vision with multi-spectral displays. The chest pack contained the all-important filtration system and water storage. He handed her the contraption. "Try this."

She took the mask and pulled it toward her face. As her forehead made contact with the upper rim, the inner screens powered on. "Oh!" She pulled it away and watched as the widescreen viewer inside faded.

"Don' be shy." Chu chided. "If you wanna glimpse out the back door, you're goin' to wanna wear that mask."

She put it close again and holding it close looked up to see the blended view of the interior from the six outer lenses. She felt Chu's hands pull the skull strap secure, adjusting snaps. Through a speaker she heard Chu. "Hold still, missy. Hold still." From behind, his practiced hands tugged and pulled on her skull and then at her waist as he snugged the enviro-pak to her abdomen.

"Ow." Her voice returned to her through speakers.

"You wanna live through it, or not?"

She hadn't noticed the mask take over and tasted a hint of bitter, metallic overtone. "Ew, what's that taste?"

"Best filtration system money can't buy."

"Say what?" Unseen in the wrap-around mask, her eyebrows wrinkled in confusion.

"Can't buy it in your world for another seventy–eighty years. Can't get a new one in my time ever." Chu offered an explanation. "We don't know how to make them anymore." Chu stopped, hand on the handle to the decon chamber. "You wanna take a look?" He glanced up at the dual stream clock. "Got about a half an hour left of daylight."

She turned to find Carl, finding the expanded width of her vision a little disorienting. He stood, smiling in hi-res, non-pixelated wide screen. She turned slowly — the wide screen was dizzying. Chu had put his mask on so much sooner than she'd expected and looked a little scary in the apparatus. With a resolve that surprised her, she moved to follow Chu who, without a word, stepped into the chamber, pointing as he shut the door, and waved to the instructions he no longer needed. Reading them, she pushed the numbered sequence, waiting for each one to turn green. Sooner than she'd expected, she felt an overpressure in her ears, and the outer door popped open.

She hadn't expected the sky to be a brilliant orange, so her first scan of the evening sky was an array of stratus clouds in banded orange and gray. That gave her the impression that she was looking through a filtered false-color rendition from colored sunglasses. Then she lowered her gaze to the valley beyond. Instead of a checkerboard of poorly irrigated fields, the wreckage of a high-rise metropolis illuminated by the orange light of a lowering sun lay in dismal harsh relief below her. The broken city conjured a half million lives cut short in a matter of minutes. Her view wandered from the broken and twisted skeletons of the high-rise areas of the city to ground level, where a two- to three-story suburb was laid out in crumbling rectangles. In some areas there were only concrete slabs where wooden structures had stood. Some of the grid pattern was obscured by drifting sand dunes.

She thought about the blast of energy that must have come over the mountains that stripped the towers below of their glass and flimsy, that blasted every living thing into organic paste or carbon dust.

"Oh, oh my —" She turned to see Chu looking out over CalStation's sandblasted handrail. "It's gone, it's...horrible!"

"You did ask to see it."

"Yes, but. Oh God!"

"How much do you think they suffered?"

"Not much, I think, unless they took shelter. A slow death would not have been easy."

Turning to see what might remain of the farmhouse, she was surprised to see no trace of it. She gasped at the sight of the metal-skinned cube. Somehow, she had expected to see some sign of the house. Looking down, a few brick pilings outlined where a house might have been in some distant past. In that flash moment of recognition, she understood the tremendous shattering blast wave that whisked away the farmhouse and blew through the city below. She tried to swallow the gorge that rose in her throat and choked. Gasping now, her eyes were flooding with tears of unfathomable sorrow and loss she couldn't fully reconcile. She turned for the door, clutching at space. Had she turned the wrong way? "Chu?"

"Here, missy."

She felt a hand steering her. She reached toward where he had been standing, blinded by tears now and blinking to clear her vision to see the doorway.

"Step in and don't remove the mask. Follow the directions." She heard Chu's mechanized voice begin to give her instructions, but she felt an overwhelming urge, a chest-heaving urge to vomit. But vomiting into an irreplaceable artifact from the future seemed somehow sacrilegious, or at least, avoidable at all costs.

Parker ripped the mask off before the door opened and threw up. She realized the mistake immediately when she took in the next breath. Coughing, she stabbed at the blinking button at the top of the enter menu. She was now in no condition to hold her breath, but she tried, shutting her eyes against the noxious air as she felt pressure of the purging nitrogen, then another blast as nitrox cleaned the chamber. When she opened her eyes a moment later, the third light was blinking green. Chu must be cycling her through from the outside. She knew she'd failed to remember to punch the second button or the third.

She depressed the release for the inner door and saw Carl's face, trying against a grin to look empathetic. Her only thought was, "Never again. No fucking way."

26

Andrew Abadon swiveled in his chair and stared out across the roof-top garden. In addition to Imbler's IT security update, he had just received two communications on related topics and needed to resolve what course of action to take. The two communications were, at best, in conflict.

MacCauley's detailed the after-action investigations from Interpol on the Paris bombing had traced the HMX to a mining company in West Virginia. But it also had links to diverse incidents across Libya and Syria from a common source in southern Italy. Peck's insertion of counterfeit misinformation had helped deflect from the Munson Mining connection.

The second, Room for Tomorrow had actually sent an invitation and prospectus to a funder's roundtable in Brussels. Underwriting financing was needed for five energy projects. He was being invited to share the conversation with some of the world's best big problem solvers: TNO, KBR Brown and Root, and NGK. Abadon Industries had been selected due to its large energy portfolio. The prospectus outlined substantial returns on investment in developing and/or established renewable energy tech research projects. He rolled his fingers on the desk in a repeating drum roll. Maybe it was time to call off the attack dogs.

A soft ping announced company. He turned back toward the double mahogany doors as they swung open. With a simple nod, Margaret announced, "Sir, your brother."

"Richard, come in. Care to share some of that Kentucky mash I brought back?"

His brother, a few years younger, and not quite as gray, was more solidly built. He was every bit as cerebral as the elder Andrew, but he'd taken the time, made the time, to let his trainer keep him in shape. Richard took the offered glass and, after a sniff, let half an ounce of the golden elixir coat his tongue. "Uhm, damn, that's good." He shivered slightly as it slid down his gullet. "What can I help with, Andrew?" He set the glass down. "I have to leave soon."

Andrew took in the subtle colorations of his Renoir across the room, gathered his thoughts. "Rich, remember that assignment we'd given Mr. Peck? Odie Peck?"

"Of course, the odious Mr. Peck." Richard almost chuckled at his own humor but swallowed the gesture. "Has he come up with something new?"

"Well, not exactly, but one of his assignments has just reached out to us. Now I don't know if maybe we were barking up the right tree at all." He shifted his gaze from the pastels in the painting to his brother's tanned face, then passed the invitation across the desk.

He continued while his brother scanned the simple announcement and special invitation to investors. "Either these guys have huge gonads, or we missed something, or Mr. Peck missed something."

"Hmmm" Richard mused while scanning the summary agenda's potential investment opportunities.

Andrew continued, "The Jutland Bank wind farm drew a lot of interest, and its list of investors was world class. It's modeled on the Thames River project. We missed those boats. Peck tells me that aside from the front money put up by RFT, the Jutland Bank project is drawing significant match money from successful Brits with excess pounds sterling. RTF's investor donor and contact list is legit." He took a small sip from his own glass and followed that with a swallow from a glass of iced water before continuing. "This new wind farm proposal for the Dakotas is right in our back yard. We need to put our face on it if possible."

"Their laminar carbon battery plant in Kentucky already has site permitting expedited for a construction start in a few months. And Peck's insiders in Lexington say the technical claims are true. Storage capacity of the initial product has more than seven times the equivalent storage by weight than traditional lead-acid and would start out at less than twice the cost per amp hour unit from the original prospectus." He paused and, with puffed cheeks, blew out a long slow exhale from between pursed lips. "And, it will provide continuing employment in coal country." He gave a knowing nod toward his brother. "Your golfing buddy, Senator

Millsap's turf. He showed me the greenlight from the Kentucky Economic Development Council. They're on board, and they have the Governor's support too."

He pointed toward the bundle of paper in his brother's hand. "Now there's that New Wave Energy farm going up in the North Pacific off Juneau." He shrugged. "DO waves ever stop in the North Pacific? We might do better to invest in that and redirect some money away from the oil lease lobby."

Richard's brow knitted in the beginnings of a frown. Andrew continued. "Now RFT's announcing a new waste-to-fuel plant outside Rio, expecting to employ almost a thousand people and turn that horrid Mount Trashmore into solid fuel pellets and fuel alcohol." He thought about the recent press coverage of the trash pickers living in squalid misery atop the city's rubbish. Images of the teeming city's favelas had dampened the mood of Rio's Olympic summer games.

Richard Abadon let the invitation fall to the table. "Are you nuts?" He opened his mouth to say more, but Andrew stopped him short.

"Yes, probably. Whether or not these guys are stealing from us? That we'll have to definitively settle with whomever Peck and MacCauley track down." He waved a data printout in the air, scanning the list quickly. "But these? These are some pretty astute investments. They are taking energy research to the field with a mind to making a profit." A wry grin began to slip sideways into a dimple. "What did Grandpa Abadon always say?"

"Ahm..." Richard shut his mouth in thought.

Andrew answered before his brother could come up with the aphorism. "The richest men in the world either inherited a fortune or they did something entirely crazy."

Richard nodded, mirror-image smile forming. "What about Peck, do we call him off?"

"Hell no, not at all!" He raised a hand in mock salute. "Cry havoc! And let slip the dogs of war!"

Richard nodded in apparent agreement. "So, you still want Reyes taken out?"

"Well, not out, but maybe out of commission. He might need another subtle message." Andrew considered his most unsavory subcontractor. "Peck's not very subtle. If he hadn't procured his explosives in Italy, the Interpol's tracking hounds might have begun to point our way." He shrugged; his face held the look of someone who's just found a sour grape in the bunch. "Who would have guessed the Italians would have procured the HMX from the same supplier as Munson Mining?"

Richard sighed, frustrated still. "I'll have a word with him as soon as I can track him down again." Richard stood and with a nod. "Thanks for the bourbon, Drew." Then left the office for his own.

27

The California sunrise burst across the eastern desert beyond Neenach's ruined skyline in bright fury. It would be hot today. Carl didn't care. He was walking in the freaking future. Not walking through a Hollywood production lot but striding in wonder through the hot dry air of August 2215. He and Chu had taken Grieg's three-wheeler cart down from the mountain house and were headed to the last of the three cloud repositories in the eastern valley. He'd been impressed by Parker's hutzpah when she went out to catch the 23rd century sunset with Chu and decided he could probably lend a hand.

At first Carl didn't understand what he was seeing, but the blinding glare through his filter mask's eyepieces dimmed as the mask adjusted the gain on the multi-spectral lenses. He increased magnification and resolved a caricature of a cityscape in the distance. Black towers stood in profile against a yellow sky sunrise, with odd structural projections that could have been interpreted as architectonic. Cantilevered platforms now hung askew in danger of losing their perch from upper stories, pointing at destinations that no longer rose to meet them. The derelict profile of the Palmdale-Lancaster-Rosamond metroplex remained in the distance. He could only imagine the architectural detail in the distant city that he could not visit. Earlier, Chu had warned as they descended the mountain's slopes in cool pre-dawn darkness, that the city could harbor renegades. Life sign indicators had been triggered by past near approaches, and on some of their early telescopic surveys, flickering light from probable campfires had been visible. Chu had visibly shuddered describing the only renegades he'd actually seen.

Their target was an Infiniti-Stratus storage site. It was one of Chu's last trips into the valley. With Grieg in Europe with Parrish, he'd begged Chu for a chance to walk on the ground in 2215. Chu reached into his thigh pack and fished a memory cube out. "Here, for you." His voice came out strangely resonant, reconstructed by the internal circuitry that included microphone, amplifier, and speakers. The complex ion-exchange filters had too many reaction layers for actual voice communication without the electronic boost. "We have plenty in the MemPak array."

In the brightening day, Carl noted that there was more greenery than he would have expected. Low scrub and drought-tolerant grasses lined the path they were taking as they wound through a former block of apartments on Neenach's south side. Shadows of the buildings in the low-angled light reached for them, painting dark strokes across the landscape ahead. He remembered the adjustment for wavelength and added sensitivity for infrared. Just in case there were animate things moving about. Carl sipped on the water tube, wetting dry lips. He'd have to watch that. Chu had said there was less than a liter capacity in the mask's water storage. It was supposed to last all day. The mission was to be short though, get to the Infiniti-Stratus site, exchange the scrubbed MemPak for a full one, look for any gems in the rough, and get the new data back to the gate for analysis.

Chu's major time killer these months had been scrolling through massive file collections, searching for material couldn't be replicated back in Wellington. Only that much reduced data would be blasted in bursts to the few comm satellites still functioning.

They came to the edge of a clearing and what Carl originally thought might have been a sports court. Patches of concrete were visible where the loose soil cover was blown clear. "Chu? What are the chances of one of those dust storms coming up?" His own voice sounded odd in his ears.

"Don't know — we don't get satellite weather here. Your time is spoiled by too much certainty."

"Hmm," was all he could muster. At the far side of the clearing, a pile of dirt proved to be recent excavation spoil removed to clear passage down a descending stairway. A pitted aluminum handrail led down at least twenty feet below the concrete ground-level roof. Carl watched as Chu inserted a five-splined key into a hole. The splines he knew were not notched like his brass house keys; they were magnetically coded. The key triggered an electronic lock and they stepped inside. Lights came on, illuminating a clean technically pure design he had not expected in the wrecked remains of the suburb he just passed through. Panel illumination in soft green emanated from guide strips along the floor, but these were completely unnecessary with the modular panels in the ceiling. Uniformly white, without any apparent heat generation, Carl thought that was a good trick.

"Can you turn down those ceiling panels?"

"Cha, no problem."

Chu turned from rummaging in his equipment bag, looked up at the ceiling and then at Carl. He shrugged, made an arcane motion at a wall panel, and the lights dimmed to half their previous output.

Carl started to work the snap buckles to his mask's restraining straps.

"NO mate!" Chu slapped his hands away from the buckles. "You want to die here, Carl? Wait! Two minutes."

Carl stood stunned. He thought they had reached some degree of safety. But looking around, he realized the room was pretty barren considering his expectations. This is only an anteroom. No secure server site would be directly accessible through an outside door. Even in the desert community that had preceded Neenach's final days, it was an area that produced dust storms.

Chu walked away toward a door flanked by two sidelights. A secured man-trap entrance with air scrubbers. That door opened into a second small hallway with a pass-card control next to the jamb. Its guts had been disemboweled, and jumper wires gave clear proof of expert tampering. Chu pushed the thumb tab on the door latch and pushed against obvious over-pressure that puffed air out into the entry hall, raising dust off the floor. Chu let the hydraulic closer push the door shut and began to undo the latches on his mask.

Carl followed suit. "Thanks back there, I thought for some reason that we were safe underground." He felt a shiver coming from the sudden cool temps in the server room.

"From lightning maybe." Chu responded with dry humor. "The atmosphere in here is scrubbed and designed to ensure that contamination can't get in. Remember that this place was secure storage for millions of users. Triple redundancy hardware, continuous backup of the system software, and more than enough power in storage to keep it all running for a week if the fuel cell was compromised."

Carl looked around, impressed. He knew server rooms; this was not his expectation. Instead of racks of black-faced drives with blinking red and green indicators, there were blunt white cylinders, five feet tall and about three feet in diameter, that reminded him of rows of overly large oil drums. He reached out and touched one of the cylinders. It was cold, not much above freezing. "Oh, it's really cold."

"Probably doesn't need to be now. When it was connected, there were a few hundred thousand files moving around at any one time." Chu walked off toward a corner console, pulled a MemPak from his shoulder bag, and set it on a silvered pad on the counter. As Carl walked up beside him, he saw Chu gesturing toward a seeming blank panel above the countertop. It came to life and began a dialog with Chu in English. Chu responded in verbal commands and an indicator on the MemPak lit green.

Carl watched mesmerized by the Chu's interrogative conversation with the machine; thrust without parry but thrust with keywords and finesse. He got access to a database he'd been searching for. His hands flew before him. Haptic enhanced fingertips weaving a sign language only the machine could read. "What are you looking for?"

"Corporate pharma records, patented techniques, specs — not only chemical processes, but the assembly line designs — operating parameters, cool room protocols, everything needed to make a particular analog of an enzyme from a poisonous Bolivian lizard that went extinct in the 2050s." The machine compiled and refined information based on filters Chu provided that minimized repetitive output and minimized the total data needed to be put on the limited storage they could take back up the mountain. Chu explained in quiet asides that the enzyme proved effective in preventing or slowing cerebral amyloid angiopathy, one of the causes of dementia.

Chu swiveled in his office chair, taking casual nips at the nail of his index finger. "We'll be ready to go in about twenty minutes."

"So, these drums?"

"The servers. What'd you think?"

Carl spread his fingers near one of the cylinders, sensing the cold sink. "I mean, what's the cold from?"

"Nitrogen, 'bout 200 Kelvin. It eats the heat."

"Cool!" Carl nodded his appreciation.

"No, cold. Really bloody cold. Capital C cold. Cold as a tax collector's heart."

Grinning at Chu's attempt at humor, Carl got caught up in Chu's uncharacteristic response. "Cold as a witch's tit."

Chu added, "In a brass bra." and laughed along with Carl. As they waited, sipping cautiously at the limited water in their masks, Chu explained his understanding of the 22nd century's server tech. Throughout the explanation, Carl was troubled. Chu just didn't do that. He'd never experienced Chu in such good humor or as spontaneously helpful. Something was bothering him.

The machine finished its task and the androgynous voice at the console intoned, "Download complete — please detach removable storage."

"Let's go, it'll be hot soon." Chu led the way toward the layered antechambers.

They donned and checked each other's masks and straps and moved to pass through the door. Room temperatures rose as each of the three isolation doors opened. A final radio comm check and they exited to the outer stair well. The last door was sucked open from outside as a surge of warm air flooded in on billowing clouds of grit. The yellow-tinted dawn had burned to a white-hot haze. Fine white dust blew into the room. An undulating complaint from blowing wind made it obvious; the weather had changed for worse. Chu pulled an unrecognizable sidearm from a side pouch, clicked off the safety and raised it to ready position.

~ ~ ~

"Quick, out!" Chu vaulted up the stairs to the ground level roof, stopped just below the surface, and rose slowly, his head protruding no higher than necessary to see above the rim of the stairwell. Puzzled but curious, Carl turned to ensure the outer door closed and climbed the stairwell. His environment mask adjusted for bright daylight, but his view was whitewashed by blowing grit. He had to adjust his earpiece to knock down the wind's growl. Blowing dust painted near white-out conditions in most directions.

Chu turned pointed two thumbs up, nodding to Carl. "You OK?" Chu leapt up with a mumbled mechanical shout and ran.

Only in running, trying to keep up, did Carl reconstruct what he heard over the howling wind as "Follow me." He fought against a sharp wind blowing left to right, pushing him forcibly off course.

Chu ran on ahead. *Damn, that guy is faster than I'd have thought!* His apparent target approached a gap between two fifty-story derelicts. Their dark masses visible above the wind-driven white-out at ground level were revealed in daylight as glassless high-rise buildings. He recognized from the hundreds of balconies that they must have been residence towers. Only one of the three sky bridges between the two remained. He had seen these towers from the mountainside in the early dawn light. He stumbled, nearly falling, and recovered.

Where's Chu? He tapped at the control that created the wire frame image and found his bright green indicator moving ahead and to the right. He damned the possibility of tripping again and caught up as Chu paused to peer around the edge of the one of the residence buildings.

Panting, wanting more water, Carl took a tentative sip, wondering how much more was in the bladder. He said. "What's going on, why are we running through the storm?" Then when there was no immediate answer. "This isn't the way we came."

Chu held up a hand and rotated his masked head back and forth across the white void ahead, apparently searching. An involuntary shiver shook the length of Carl's spine. He hoped his heavy breathing wasn't being magnified by the speakers. Instantly, the sound of his amplified breathing did diminish. Chu held his gloved fist up in stop mode and said through the comm, "Get ready to run." Then with his fingers only, counted down — three, two, one — and bolted from the cover of the building.

It seemed to Carl as he followed that the gusts had diminished, but the near whiteout persisted. Another block of buildings appeared through the haze and, finally, a large field of barren pavement. Ahead, Chu seemed to jump for no reason, but then he saw the double-row of wheel stops and slowed to avoid tripping. A parking lot! He wondered, as he tried to keep an eye out ahead and now down for trip hazards, if he fell would Chu stop and look for him. Then he fell. He had jumped over another of the low concrete wheel stops and landed a heel on the wheel stop for the facing row. Landing on knees and elbows at speed, he slid to a stop. "Fuck!"

He got to all fours, mentally feeling for what hurt. He determined that he'd get by with bruises in places that would be sore as hell tomorrow. He called into the mike. "I can't see shit! Slow down, I fell and can't see where you're going."

Over the internal speakers, Chu's granulated voice said, "Press the small button under your right ear."

He had already done this several minutes ago and what seemed like a mile ago. He tapped it again and the wire frame filled in. Shaded planes appeared on the inside of his dusty lenses. A 3D solid world developed. A small green ball with a black three in near center of vision bobbed up and down, and slightly left to right. That was Chu. A dark shape moving in from the left seemed to be closing in on Chu. *Run!* He saw the next row of wheel stops and handily jumped over them in stride. His heart was pounding now. He gave a second or two of thought to Tom Corliss when he saw a black shape. A simple dark square was moving low and fast, following Chu's green number three ball. He was bringing up the rear! *Shit, whatever that black thing is, it's between us.* Another black square was closing in from the right, moving parallel with the first, and still between him and Chu.

Sweat trickled from his forehead and down his nose, impossible to scratch it away. *Don't even think about lifting the mask!* They were approaching another building. *How many of those shapes had there been? Three? Four?* He heard a vaguely canine yelp through the earpiece. The black square stopped moving. Abruptly, he came to a stop behind Chu who was crouching out of the wind behind a low wall. A large dog—no, not a dog—was spasming in death throes a few feet away. "What's that?" He pointed toward the creature; its legs jerked in the last throes of spinal death. Chu didn't answer but was tapping on the back of his left hand with the fingers of his right glove. Chu's head was bowed. His crouch, one knee down, his elbow resting on the other, gave the impression of a disciple in prayer. Carl realized he was actually conducting a search on his smarter-than-he'd-realized environment mask.

The noise of the wind howling through gaping holes in the lost city filled his mind, but he leaned close and shouted. "What are you looking for?" The sound of his voice was loud in his own ears.

"Please don't shout. The mikes will do the work for you. Just don't puke on them. They're all hell to get clean again." After a wind roar-filled pause, "You've been shouting since we left the data center. Bleeding rude."

"What are you looking for, the way back?"

"No, scavengers."

"Shit!" Then, "Animal or human?"

"Either, both! Doesn't make any difference. The renegades are animals too."

"Why weren't we looking for them on the way in."

"I was, but they're more dangerous when they can hide, or think they can hide, in the cover of the dust."

"Oh! OK. Where's the other dog thing?"

"Naffed off! Don't care—a three-legged dog won't be back!"

Carl was impressed. One of the dog things was dead, the other lamed, and in just above zilch visibility. "Do you know where we are?" He was about to add "or where the car is?"

"Cha, shyte, stay close." He was off again into the pale void.

Carl jumped after him, leaning into the wind. His right knee was complaining now. The forward dive into the parking lot was paying out its dividends. Nearing the edge of the high-rise district, the wind was stronger. Chu took a left around a long, low building. It seemed to Carl to be the loading dock side of a commercial or retail building except the doorways above the recessed ramps were circular. They passed a machine recognizable as a trash compactor. Much of its thinner sheet metal had been sandblasted to wafers, but the shape was obvious. Carl pulled again on the water tube and got the bubbly slurp of an empty container. *Hell, I was wanting that to not go dry yet.* Chu pulled further ahead. The green circle was fading.

Adding to Carl's rising panic, the signal from Chu's mask disappeared. He paused and tried to point his head into the white haze where he thought it was last seen. He pushed again at the button under his right ear. The visible band dropped away, and on the darkened screen the green circle reappeared. A minimalist wire frame model of the ground ahead appeared. He heard a canine yelping in pain that stopped with fatal finality. Chu was holding off more of the dog things.

He rushed toward the green circle, taking higher than necessary steps, afraid of tripping on some detail the ground form generator might miss. To his relief, the position indicator reappeared and blended into the form of a orange square, a box really. He careened at almost full run into the three-wheeler. Carl heard overtones of a chuckle in Chu's voice. "Congratulations, you found the ride home. Get in, shut the door." The wind noise dampened considerably.

A snarling disturbance rose above the dry maelstrom outside. Two dark shapes circled the vehicle. The growling reminded Carl of the high-pitched complaint of coyotes, but these creatures were no coyotes. Closer in size to German shepherds, they had oversized shoulders and undersized hips. More hyena-like in profile, their faces were pure malevolence. One of them bumped the vehicle's door, sniffing at the handle pull, then tested it with his teeth. The other jumped to the short hood and presented its bared snarling canines through the windshield. Something bumped into the little cart from the back.

Carl turned to see a humanoid shape raising a club-like object. Chu pushed the speed control to reverse and pumped it one notch. The creature on the hood yelped and fell off. The human thing behind fell beneath the wheels and cried out a yell and a guttural curse. Chu pressed on the horn button and the two canine shapes sped off into the white. The little cart sped forward now, bumping roughly over the fallen human attacker.

Carl let the trapped air in his lungs slowly escape. He was now aware of the bitter metallic taste of adrenaline on his tongue. "Son of a bitch! That was...interesting!"

After a moment, Chu's mask nodded. His face was hidden, but the voice had humor in it. "Manno, who said, 'May you live in interesting times'?"

"Damned if I know."

"Well, now that you've seen both, what's your favorite — 2215 or 2021?"

"Home. Get us home. Just get us the hell out of this valley."

"You do still have that MemPak?"

Carl felt for the lump in his shoulder bag. "Yes."

"When you get back, see who is most likely to benefit from the PV shrink wrap."

"Is THAT all you got to say about the weather? About whatever those things were? About that man you probably killed back there? Christ almighty!"

"Should have warned you; it's fairly common." The mask turned toward him, and the metallic voice of Chu's mask said unconvincingly, "Not usually as bad as today though; sorry."

"Why didn't we park six or seven blocks closer?"

"Scavengers, that other kind. The one with the club."

Carl now fully appreciated the daily dangers Chu and Grieg were exposed to on any given data retrieval. The cart moved upwards away from the ruined city, toward safety. At each switchback, Carl eyed the malevolent, roiling brown river of dust flowing south along the valley floor. He had never really experienced anything worse than a Santa Ana windstorm before, and they hadn't harbored marauders with hunting pack canines.

After his own breathing had steadied, Chu turned to Carl. "Those scavengers will kill for a sack of rations, and the contents of that server room would just be wrecked for the simple joyous luddite hell of it. We can't afford to park near enough to let anyone know where we're going. And that spot was shielded on three sides."

"What about that guy?" Carl's thoughts were rampaging through the implications of living in a toxic, lawless Mad Max version of the wild, wild west. "How do you so blithely drive away from a severely injured man? What if that was Tom Corliss?"

"Tom is long dead; don't burden yourself with his welfare." Chu said. "That thing back there probably shared Tom's meat with his dogs." After the three-wheeler settled after a series of hard bumps, he added, "We've seen that guy, or, some guy out with two or three hunting dogs before. Glad he's down. Hope he's out."

"You think those were his dogs? Not wild dogs?"

"Cha. His dogs, but they are doberman smart, pit bull vicious, and coyote cunning. I shoot them on sight. If that guy is seriously hurt, his dogs will probly eat him."

Carl closed his eyes in thought but wore a studied frown beneath his mask most of the way back up the mountain. A formation of house-sized boulders passed on the left, and Carl knew they were almost back to the portal. "Chu, can I ask you something?"

Chu turned toward him. "Back in the Chick-a-Saw server room, you were, well, it seemed like you weren't yourself." He let the question float. "I'd never seen you like that. Goofy almost."

"Nothing for it, manno. Sometimes I remember the good times I had with Noel. We used to have some good times together down in that server room." They bumped along the track for a few more hundred yards before Chu added. "We were more than mates, right? I mean, you knew we were a couple, and ... now that's over. He's gone. I'm not."

"Sorry, Chu, I —"

"S'all right."

28

Carl cycled through the decontamination chamber following the instructions Chu had given him and reset all the controls and closed the outer door for Chu's entrance. While Chu was climbing in, Carl looked by long habit for his cell phone. It buzzed again as he pulled it out of his driving jacket. The messenger line held a simple text message from Joan Brayton. [Parrish landed at LAX Developments return ofc ASAP F2F here only, U2 no F2F in car].

He studied the line of txt, deciphering, then understanding. No talking in the car. His mind raced with implications, but he had other more urgent concerns after four hours in the hostile environment outside the time station. Two minutes later, Chu emerged from the decon chamber to find Carl stepping from foot to foot in impatient exasperation, his eyebrows knitted in question. "The dunny is right there, thought you knew."

Carl stepped out of the toilet two minutes later. "We need to get back. I had a message on the phone. I, we, need to get back to Anaheim."

"I need to do some things first. Can it wait?"

"Brayton said ASAP. She never says ASAP; she usually assumes it, so when she says it, that's important."

"OK, can you tell me what ASAP means?"

"Jeez, yeah, ASAP means 'as soon as possible.' We need to go now. This time of day, we've got over two hours in traffic."

"One second, maybe three." Chu pulled his transfer MemPaks and set them on the console's I/O pad. A red square appeared around the perimeter of each MemPak's base. One by one the red indicators turned green. Somewhere in the depths of the desk console a cooling fan spun up. The console came to life, and Chu gave it appropriate commands in hand signal syntax that Carl found recognizable, yet not understandable. He watched, fascinated, as Chu scanned a few master menus and stopped the scroll, pulling one of the floating sheets into focus. "Fuck the world, let me off!"

"What is it, Chu? I really need to get back. Do you want to stay and work?"

"Manno, idle it. Maybe ten seconds more." Chu barked a few rapid commands; the sheet crumpled. Another one rose and focused, then folded itself and slid off the screen to the left. "OK, let's go, I just sent the summary in an email to your phone."

Carl was already heading toward the gate room's portal. He stopped and turned. "That's not secure. At least we don't trust it from week to week."

"If that email is intercepted, there's no one on this planet except Parker and Grieg that would believe a word of it. For anyone else, it's jabber."

In his pocket, Carl's phone buzzed. He took the steps down two at a time and had the car's engine purring when Chu finally joined him.

Chu buckled in and turned to Carl. "Well, what is it? What's so important? Car's not secure, is it?"

"Probably not. Joan Brayton said not to talk in the car. But I guess the same levels of incomprehension would apply to third parties listening to us talk about Neenach. Cause it's a backwater with no water. No city, no tech center." He put the car in gear and guided it out toward the roadway, conscious of leaving minimal tracks in the gravel.

"There's been a change. Johannesburg is online." Chu let the message float, waiting for Carl to piece the two statements together.

"A CHANGE!" Carl waved his hands in celebration in the limited motion allowed by the Porsche's driver's seat. "Johannesburg's online? Does that mean that you can go from gate to gate?"

"I don't know — probably no. It does mean the area is not as hot as it was, or the circuits wouldn't be working. Someone is in the neighborhood generating power. If we've managed to change anything, there must have been fewer detonations over South Africa. That's all I could pull off the system because you were in such a skittering blind-bat hurry."

"Crap, sorry!" He took the first two turns down the mountain at speed but then slowed. "You want to go back?"

"Nah, let's get on down. There's detail in that email. I just have to unpack it."

More than two hours later, with rush hour's inevitable soul-crushing delays, the red Porsche 930 slipped onto the I-5 southbound and began to make hash of the light traffic heading toward Los Angeles. As they passed through Glendale, a black Mustang joined them and kept pace a thousand yards behind, sure in its more cautious approach to thickening traffic because the tracking signal on the surveillance GPS was doing a great job of locating the Porsche.

Almost an hour later and only a few minutes apart, the two cars reached the Anaheim office of Room for Tomorrow, but only one of the cars pulled into the executive parking area. The mustang paused at the curb for more than a brief moment, and with its signature low growl obscured by background traffic noise, pulled away, slowly and unseen.

~ ~ ~

Chu and Carl joined Parker and Grieg in her office's outer waiting room. Parker called out to them, "Come on in, we have news." Joan Brayton followed them in and closed the door.

The inner waiting room had a small oval table and couches. A wet bar contributed to its occasional use as a lunch nook. Glasses of water waited next to a small iced bin. "Help yourselves." She grabbed one and kept walking. "But follow me. We have a problem." She turned on her black block-heeled oxfords and headed for the small Japanese garden on the back side of the building. She'd had it installed on the roof, shortly after taking title to the building, and it was beginning to look like it had been there for decades. Lichens and mosses were taking over in the microclimate afforded by misters. A glass wall blocked street noise but not the views and offered a comfortable retreat, serenity in the middle of a commercial hub. Parker didn't feel serene. When they were all outside, she took a deep breath and exhaled slowly, pushing the tension down and out with a purposeful, eyes-closed shove of her hands down toward raked gravel. She walked slowly to the tempered glass wall at the edge of the rooftop retreat.

She turned to face them, a changed person. "Everyone, I had a visit from Joanie this afternoon." Heads canted, or shared questioning glances. "We're being surveilled again. Spied on! I don't like it." She took another deep breath. "Carl, can you do something?"

He glanced quickly toward Joanie. "Brayton and Eastberg can get on it immediately. We'll deal with it, hon."

Brayton nodded an affirmative.

Parker shot him a piercing glance that said *I'm not 'hon' right now* and began to pace, trailing a hand along the barrier. "Joanie, your report today seems pretty conclusive that somebody with resources is keeping an eye on us. My car is bugged. Carl, your car is bugged. Whatever you guys talked about on the way down here is now in someone else's database. Our houses are bugged, again." She stopped pacing for just a second to recalibrate. "Joanie, you might as well put your house on the list too, if someone's been trying again to get information out of our finance and accounting server." She shifted her glance from Brayton back to Carl.

Carl took the news calmly. Although several years younger, he was slower to react and more purposeful in reaction. "Park, after some news from NZ and the 23rd that no one in the here and now will understand, we small-talked about weather and traffic or listened to jazz. I don't trust the car anymore if it's been out of my sight." He looked toward Chu who nodded affirmation. "So, Joanie, we've been hit again?"

"Yeah, pisses me off that our security has been breached." Joan Brayton's smile was thin-lipped, serious, on the edge of a grimace. "But the left tracks." She shrugged. "That may work to our benefit. I think I can live with that."

Carl's head bobbed in affirmation.

Parker stared at them both, incredulous. "What? What the fuck are we going to do if the Abadons, or Chained Industries, or the Russian mafia is sleuthing our door, or the freaking IRS? They won't be happy either."

"No sweat, Park. I can put up another set of false fronts for tonight, and when they come knocking, we'll let them in to see the other books?" Carl offered an impish *Non mea culpa* shrug.

"Other books?" Parker sounded surprised.

"Yeah, Harriet Eastberg is a real wizard."

Parker's eyes narrowed in question.

In answer, he offered, "Like any good corporation based on felonious fiscal arrangements, we've been working under two sets of books since about day five. I must have reminded you months ago."

Parker waved a dismissive finger in the air. "I thought you were doing that with your spreadsheet thingy."

He looked at her with comingled affection and irritation. When she was upset, seriously upset, she seemed to lose IQ points. "Parker, you know we've brought Harriet into the inner circle. Otherwise I'd be thinking about retirement in New Zealand right now."

"New Zealand?" Parker was seriously irritated now. "Really? REALLY? Crikey O'Reilly!"

Chu had been standing there, in the inner circle up at the farmhouse, but a silent witness in the corporate HQ. He stifled a snorkling giggle. The euphemism, one of Chu's minor blasphemes, took Parker back to the portal in the farmhouse and struck her as suddenly too funny.

She couldn't help letting a grin erase the scowl she'd worn since she'd heard the news about being bugged again. To Chu she muttered an aside. "Jeez, glad you're developing a sense of humor." She pivoted to Carl. "OK but do something fast. I don't like being in a fishbowl."

"Hon, this is what I do. Joanie? These are all external to the building, right? No sign of anything on campus here?"

Joan Brayton relaxed her rigid jaw, taking stock of all she knew. "All the bugs were found on private vehicles and residences. There was even one mounted on my apartment's cable routing box to snoop my telecom. Remember when I began looking and found them in your houses and cars? I don't know of any reason to suspect the farmhouse yet, but it should be scanned."

Chu and Grieg looked at each other. Chu said, "We can surveil in our time; we have motion triggered multi-spectral vid. We have nothing on the farmhouse but a simple proximity alert in the house itself."

Joanie volunteered. "I'll scan the farmhouse." She pulled her phone and began to enter the scan on her to-do list. Looking up at Parker, she added, "I'll redo the beach house first. Tomorrow morning."

Carl said. "They can knock on the door, but unless they are in bed with us, or looking over our shoulders when we're online, there's not much chance of—"

"Carl!" Her voice was soft, but sincere in its pleading.

"What?"

"Just make me believe I'm safe, that you're safe." She expanded her glance to take in Chu, Grieg and Joanie. "That we're all safe."

"I can do that. Joanie's already hardening the RFT server wall, right?"

"Yes," Brayton paused, "Yes, that's being done as we speak."

Parker still wasn't mollified. "Yes, but the bugs? My car, again?"

Grieg spoke up. "Fleet cars."

Everyone turned to stare at him, wondering at the apparent non-sequitur. He continued his calculation. "How many employees work for you?" He had turned to look over the garden's hedge toward the parking lot.

Parker answered, "Right now, about 225; next month we'll be pushing 250."

"OK," Grieg recalculated, "How many have been with you for over a year?"

"I don't know, about thirty or so I guess."

"That'll do." He was met with three puzzled stares. "You need ready, dependable transportation, right?" His eyebrows raised above a conspiratorial smirk. "That you can have private conversations in? You are also a leading proponent of environmental change." He had moved to the edge of the roof garden and now motioned to the parking lot below them. "Your parking lot is only a little more of a liberal statement than any used car lot. Everything from an aging Prius and two hybrids to an ancient eighties panel lorry that probably gets less than thirty liters per kilometer, or, er, ten miles per gallon." Studying it further, he added, "But on whole, it's definitely a statement reflecting who works for you."

"And....?" Chu drew him out.

"So, you buy a new Prius for the use of everyone with more than six months' time in service. Or a Leaf or a Volt, you get the idea. I'd bet you could get a deal on them if you paint them in RFT fleet colors and put on appropriate logos for street presence. And you need to get rid of your personal parking spaces. If you need a car, take a car. It solves a couple of problems. If someone has infiltrated us, they don't get one yet. You have time to ferret them out." He was formulating the plan on the fly, seeing the difficulties. "Number the spaces. When someone arrives in their company car du jour, they put the key fob in a bin, or hook, or something with that number. If you need to go somewhere, you grab a numbered key and go to that space and drive off. No one leaves personal stuff in any of the cars."

Carl objected. "But I just ordered a new Tesla Roadster!"

Grieg pushed on. "For the electrics, you put in ample charging stations. Put 'em in the preferred parking locations. Great PR too."

"But—" Carl tried to cut him off, but Grieg persisted.

"But nothing. If you tell the employees that at four or five years, they can take one of the fleet cars as a bonus, you're gonna have one hell of a loyal workforce."

Chu eyed Joan Brayton and asked, "Can you put stealth optics and silent alarms in the fleet?"

"Yeah, I guess we should do that; and to your sporty cars too," she said, looking at Parker. "Trackers wouldn't hurt either, in case we have reason to look for an insider at some point."

She nodded assent, thinking about a thirty-odd fleet of new electrics as a standard commuting car. "That will take a little fun out of the ride home. Can I get a Tesla?"

Brayton cut the dream. "Well, nothing more than the basic Model 3. Getting you a Model X would eliminate the randomness of the vehicle you'd pick to drive home. Unless you want to buy thirty upscale Teslas." No one took that suggestion seriously. Seeing her bosses crestfallen pout, she added, "Maybe we could get a half dozen or so of the long range Model 3s." She added, "for senior management."

Parker huffed a frustrated breath and looked at her toes, lost in car pool mathematics. Carl scanned the urban horizon beyond the garden. "And we hire walking security. Ex-SEAL, Delta Force, or something like that."

They began to take the planning for enhanced security into the weeds. Chu had pulled his cell phone and decrypted the email he'd sent himself. "Bleedin' 'ell." His wide-eyed stare at the screen said there was more.

The small group turned on him, interrupted by the vehemence in the outburst. His California-tanned Asian complexion had paled. Pupils dilated in an inward stare.

Two or three *whats* sounded in unison.

Chu looked up from the phone, but at no one in particular, then turned to Grieg. "I, we have to get back to the farm." He caught Parker's eye. "Can we use the Jeep?"

Grieg reached for his arm, offering comfort. "You look sick, I'll drive."

Parker persisted, "Chu, what's the matter? Of course you can."

"Paradox." He turned to focus on her directly. "Temporal paradox."

29

Meg's entrance into Parker's office was slow, soundless, and respectful of the concentration her boss sought in what Parker called her R&D sessions. The soft mint green and Williamsburg blue retreat she had created in her corner of the eighth floor was restful when the glass wall was darkened. Beyond those coated windows, late afternoon sunshine had turned the Los Angeles basin's air a malevolent brown going on ochre. A typical frame for the setting sun as the evening's commuters poured another million cubic feet of exhaust gasses into the unmoving mix. In the back of her mind was the knowledge that fires in the lower Sierras were adding to the orange brew. Fires of unheard of frequency, and extent.

Parker's temporary solution was to not look out her windows at it. Her attention was fixed on her twin monitors, engrossed in reports of energy blackouts across the Caucasus and Balkan states. "Anika Kovačić."

Meg halted in mid-step, rocking to an unbalanced stop. "Who?"

"Anika Kovačić is a principal in the state-owned power system in Croatia. I want you to reach out, set up an appointment for Carl and me to talk about an installation of enhanced glass PV panels." She swiveled in her chair, away from her array of monitors, and faced Meg who had regained her balance and stood expectantly halfway from door to desk. "Her husband, Petar Kovačić, is assistant to their equivalent of Secretary of State." She steepled her forefingers absentmindedly.

"Meg, Carl and I were talking the other night, and it occurred to us that stabilization in the Balkans is several generations down the road. I think the phrase for ongoing racial tension is "people gonna hate" or something like that, but the hate comes from want, and shortage, and ignorance."

"Haters gonna hate, Parker. It's haters gonna hate, or 'all you haters gonna hate'." She smiled in apparent self-satisfaction. "White girl rap from, I don't know 2013 or maybe 14, and more recently the last presidential campaign."

"True enough, whatever." She turned back to her screens. One displayed a highly detailed aerial photo mosaic at sharper-than-Google-resolution of mountainous terrain and a coastline sprinkled by tiny orange rooftop rectangles. The other had a patchwork of spreadsheets, reports, and email windows overlaying one another. "This is western Croatia, on the Adriatic, a beautiful vacation spot for enough of the year that it has become a major source of income for the country. From space you can't tell that two decades ago armed gunmen were herding up Serb families, loading them into buses to be physically removed, usually to refugee camps, sometimes to burial pits. From space it's just a lovely place to spend a month's salary in two weeks, take in the spas, and work on a tan."

Meg waited, aware that Parker would throw out a factoid as if a random page of history had been evicted from an encyclopedia. But the factoids were always followed up by another idea for a project. So she waited.

Parker had a touch pad stylus balanced between forefinger and middle finger and had been tapping its soft end rapidly on the edge of her desk. The tapping stopped, indicating a decision point. "Meg, if we can help the Croatians produce their own energy and get unplugged from the Russian pipelines, not only will they be able to lower the cost of production and the cost of living, but we reduce the impulse for conflict."

"Yes, but the ethnic structure of the Balkans — "

" — is screwed up, I know." Parker interrupted. The tapping of her stylus picked up again, up tempo. "Priests, rabbis and imams have been messing with these people's minds for generations, but they've always had the fabric of generational poverty as a context to weave their agendas onto. If we can get something going over there that boosts prosperity, why, I think in the long game you're going to be less likely to put on a bomb vest if you're building car parts, or flat carbon batteries, or whatever." The stylus stopped, poised.

"Carbon? There's coal there?"

"The hell with coal; there's a huge oil-fired plant on an offshore island that typically throws its carbon into the wind. We scrub it for molecular carbon, which is what we need anyway for laminar carbon." We can get the carbon from the oil plant's new scrubbers for the design life of that plant, then we'll no longer need that plant." She turned to face Meg, waiting to see if the dropped hint about the 'new scrubbers' was taken.

Meg nodded, following Parker's lead and smiled. The stylus, dropped, rolled to a stop against a short stack of printouts. She offered. "Flat carbon batteries for storage of solar with carbon scrubbed from an oil-fired plant?"

"We can hope, Plan A anyway. The Adriatic coast has tremendous, well, not tremendous — " She turned to the map screen, toggled a button, and the map changed to Eastern Europe in gradations of brown in the south to green and blue in the north. She pointed to the eastern Adriatic. "See this brown band up the coast here? Most of these regions only a short distance inland have about the same mediocre solar potential as southern Canada, or maybe Nebraska, but, you know, cloud cover." Her voice trailed off.

Meg was, in fact, extremely knowledgeable about the climactic variation in solar efficiencies. She organized the raw data for many of the literature searches conducted in the two executive offices. When Parker or Carl got another brilliant idea, she had become at least passingly familiar with most or the cutting-edge field research.

Parker continued, picked up a remote controller, and brought a wall screen to life. The 84-inch super-max display copied the map screen, and its cursor stroked the medium-brown tones along Croatia's coastline. "This section of islands and coastal communities has the same solar potential as southern Italy, or central Spain. They just need to make solar more than an adjunct to a petro plant, more than a feel-good supplement. And they shouldn't have to get it from China."

Meg got it. "Flat carbon batteries for large scale storage. And if we're not yet ready to use Croatia as a battery test bed, then hot salt storage?"

"Yes!" She fist pumped the air. Parker's tone was ebullient, barely contained, her grin infectious. "And you're probably right about the hot salt storage. There's that researcher in Florida building thermal salt pilot projects. Let's see if he wants to go upscale; toss some grant money his way or maybe find angel funding."

Meg closed her lids in thought, remembering the name. "Amratiputhi, or Amratputhi. I can find him." Do you want me to send you his contact card?" She waited to see if that was the end of the creative burst that had called her into her boss's office.

Parker pursed her lips, as if her brainstorming had snagged on a detail. She leaned over her keypad and pulled up a calendar full of text entries and colored time blocks. "Damn, I probably can't go. But Carl could. See if you can find something in a week or two for Carl that matches any openings Anika Kovačić. And Amratiputhi, pay him any amount of money you think decent to get him to go along. We aren't going to want to ship those hot salt systems over there. We're going to want to build them over there. Plenty of sulfur being scrubbed off those oil boilers now to put to good use."

Meg stood, waiting, her shared smile still in place, in admiration. She caught Parker's glance, just before she turned to go. "Parker, there's something I wanted to—"

"No, I can't go. It would be fun. But I'm heading to Brussels. Meetings with some EU types and a few less dignified folk that Carl has contacts for. Stiff-lipped bunch, but we need them. Real cash investors, not just world banker rich, but Cayman rich. If these meetings are successful, we might be able to afford to stop stealing money from the robber barons of the world."

"That's not the thing, Park. I was just thinking of us, a year and a half ago—finally free from corporate law and meaningless litigation, defending idiots. We had just hit the wall representing the same scumbags as they broke up with their wives so they could marry younger wives. The Law sucked." She raised her palms outward, in an all-encompassing gesture. "Now look at us."

~ ~ ~

Mouth agape, Grieg stared at Chu. "You're serious? Hickok isn't in charge?" You sure?"

Chu gripped the wheel harder as he took a hard right on the gravel approach to the farmhouse. "The message was from a J.

259

Chesterfield who never heard of Hickock." He was driving a little too fast for the road and had only recently learned to drive the Jeep, but he felt urgent need to put in a call to NZ to check on any changes.

"Oo, da bleedin' 'ell is Chedafeel?" Disbelief and the reference to home had jarred Grieg back into his 23rd century patois.

"Chesterfield," Chu peered forward into an imaginary vanishing point even as his subconscious helped him keep the front of Parker's Jeep between the curving ditches. As the road's multiple switchbacks mounted the slope, the terrain on either side alternated between brush and boulder, to inadequate guard rail or blank space. "Chesterfield is the new director." He frowned as he remembered the text message, forwarded from the time station's comm unit. "Chesterfield's signature mark was on the transmission in response to my last report to Hickock." He blew out a puff of air in exasperation. "Section Chief Chesterfield wanted to know who the bleeding hell was Hickock and was something wrong with the equipment that we haven't reported in three weeks."

He had been trying to stretch his memory to anyone he might have known at the service by the name of Chester or Chesterfield and kept coming up zed. The drive out of the city had been harrowing, his nerves were shattered by the seemingly incomprehensible need for other drivers to make lane changes. This was only his second attempt at driving a vehicle in traffic that required driver attention. He'd nearly caromed off the fenders of fellow commuters far too often. They had waited until well after evening drive time, and still his skills at piloting a driver-guided vehicle had been challenging. If he had not had the practice of the three-wheel utility truck that took them down into Neenach, he would not even have tried. In the growing darkness of evening, they passed a familiar boulder the size of a semi-tractor, and he knew they were approaching the farmhouse.

Chu said speculatively, "You know, for a second I wondered if CalStation would still be there or if we would find an empty bedroom in the farmhouse." He stole a glance sideways to find Grieg still staring at him. Grieg's face showed new concern for the prospect that their return ticket couldn't be punched. "Don't worry, after my first mini-panic, it came to me that if the time station was gone, it couldn't have forwarded the message."

Grieg just grunted. "So? Hickock doesn't exist anymore?"

"He might still exist, but not in a career track that brought him to the top of the Time Service."

Grieg gripped the passenger handle above the window as Chu took a curve a little too wide and his breath stopping view into a darkening abyss outside his window made him look away. "What if we killed the bloke?"

"We couldn't have. We just may have affected something, some outcome. Hell, Grieg. He might never have been born." He thought about that for a second and reconsidered. "For all our activity, we really haven't done very much toward averting the Last Day. But the general line of history still happened; the Time Service is still there." He nodded ahead with his chin. "Our time station is just ahead." The gesture went unnoticed in the dim green glow of the dash.

Grieg peered ahead into the shifting cone of head-lighted dirt track. The now familiar hills and their path up the mountain were far less ominous now than in the 23rd. "Chu, what if we do something here in the 21st that bolluxes us. Do we just stop being just because we never were?" Grieg lifted his free left hand and snapped his fingers. "Poof, I look over and there's no one driving?"

"I don't know." He allowed a small barking laugh. "Be ready to catch the wheel."

"What?" Grieg's hand, still raised, drifted left toward the steering wheel.

"Kidding." Chu wondered in childlike free association as giddy surreality stepped in to replace his normally staid programmed existence. So much of his life had been laid out with so few choices. Swim or drown, school or poverty, career or marriage, Time Service or North Island Provincial Police, and most recently, remain in a fracturing relationship with Noel or try to fix the intolerable ruin of humanity. An unnatural pull tugged at the corners of his mouth. He found himself grinning—tight-lipped still, but grinning. He swallowed another laugh as it attempted to bubble unbidden. The sound that came out was more of a grunt.

"What? You say sometin?" Concern flavored Grieg's question.

"Nah"

"You sure?"

"Nah." This time he let the giggle erupt. The urge to laugh in the face of despair was proving difficult to tamp down. "Tell you this, manno. We going to go up those stairs, step down the hall, and back into our filthy, dying, fucked-up world, and every time I step over that threshold, each time I get that punch of nausea from the time shift. It's like our world is welcoming us back into its poisoned grip."

"Each time we come out here." He waved a hand into the cone of light ahead, implying that the often-polluted air of 21st Southern California was far preferable. "I hate going back. I have a visceral hatred of going back in there." His train of thought went through the two doors and passed out into a toxic hell storm. "But there's something else about that message."

After a significant pause that made Chu think he hadn't actually said it out loud, Grieg asked in obvious expectation, "OK, what?"

"The message came from Pretoria!" Chu waited, let the world-altering truth of the existence of the South Africa Station branch of the Time Service to sink in. "Bleedin' Pretoria, manno! Pretoria is alive!"

They slowed as the Jeep came around the final turn, its headlights bringing the sun-bleached clapboard of the house into stark contrast with the hillsides in backdrop. Green had overtaken brown in the last growing season. It almost had a sense of normalcy to it. Both of them looked at the humble derelict in anticipation.

Grieg simply said, "Shyte."

30

Beyond the wall of insulated low E-glass, another midday sun burned the foliage in Parker's professionally landscaped cliff-side garden in blistering semi-tropical heat. Having oceanfront property meant cool refreshing breezes only if those breezes had passed over the chilly Pacific current. The hot winds blowing across the Pasadena Mountains today were dangerous to fatal, depending on how much air conditioning the poor could afford. Across the Los Angeles Basin, homeless sought shelter in the relative cool comfort of drainage pipes.

Glass-smooth, long frequency ocean swells rolled in near the shoreline, all signs of other winds sheared flat. Often, in comfortable weather, Parker kept a window or two open to allow the soothing white noise of the breakers below to fill the house. Today's weather was not comfortable.

"Carl?" Parker stirred, moving a leg from under his. "You awake?"

She felt him stir, either in response to her movement or the question. "Carl?"

"Yeah, I'm here." He shook his head to free cobwebs.

She needed to talk, and the need for privacy required the masking susurration of surf. "Come out to the patio." Abruptly, she stood and led the way onto her narrow perch above the cliff face leaving the sliding door open behind her. She tapped a button on the jamb, and a sun-proof canvas roof rolled out over the seating area.

They got comfortable in either end of her ocean-facing patio couch, sitting again with legs entwined. Earlier, they had fallen asleep in the quiet of a late Saturday morning, he on the last of his Paris bombing pain killers and she on the somnolent effects of a half carafe of good wine. She wouldn't let him drink on oxycodone.

"Doctor's orders," she said in stern mockery of a physician reading the riot act.

He'd seemed dutifully subdued. Turning and facing the ocean, she said, "We need to stop fucking around."

He breathed in sharply and sat up. "You want to stop—" He paused. "—sleeping together?" Confusion knitted his brows. "What are you saying, Park?" He withdrew slightly, drawing one of his extended legs under him and sat up a little more erect, wincing at residual pain not blocked by the mid-level painkiller.

"No. Sweetheart." Her damp eyes regarded him, his tousled near perfection. "I especially do not want to stop sleeping with you, or playing with you, or living with you." She was having a waking dream of sorts, possibly brought on by the '09 Barbaresco, that was taking on clarity, disturbing clarity.

"Our trip to Florida? Chu and Grieg in Latvia?" She let the two dangerous and probably foolish missions float in the air, gather substance. "What the hell! We could have killed people, as careful as we were to try not to kill people and not getting caught." Into the following silence, she added, "Actually, there were those three heat-related deaths in West Palm, no AC."

"We couldn't have known."

"We SHOULD have known." She hadn't meant to raise her voice. She pushed the heels of her palms into her eye sockets, attempting to clear out the wine. She held her hands out, palms up as if weighing, and alternately raised and lowered them. "Cause, effect. Cause, effect. I'm sorry. I'm not mad at you, just—" She reached for his other leg and pulled it back toward her, caressing the fine blonde hair on his shin. Then, cupping the muscled calf, she subconsciously kneaded it with her palm. "I think I was trying to re-create the Monkey Wrench Gang in a modern context. But that's bullshit. It's just protest."

"But—"

She held up a hand, stopping his retort. "These days, protest only lasts for one or two news cycles unless they catch the perps, and God I hope they never get near us." Visions of a steel-doored concrete cell caused her to blink and shake her head. "Damn, I shouldn't have had the third glass." She let her eyes scan the distant horizon. "We have a pretty good bank account now. Meg just showed me the numbers yesterday for RFT. We have over 350 million in unencumbered assets. More keeps coming in every day, every hour, as the phone banks, Facebook ads, Tweets, Snapchats, and mailouts do their magic." She laughed a short puffing, "Hah! Only you and the devil know what's in VSF's coffers."

She waited, watched as a grin slid sideways into his left cheek. "It's a lot."

"Yeah, I don't think I want to know exactly." Her tone wasn't convincing.

"Almost a half billion. Again, unencumbered."

"Switzerland?"

"Grand Cayman."

"OK, so between our public and private assets, we have a lot of goddamned money!" She found herself suppressing silent laughter. "So, there's no reason to go around pulling stunts." She let her tipsy tendency towards mirth subside. "No more blowing stuff up."

Carl canted his head sideways, taking it in. "Well, that would improve our life expectancy." He stretched his shoulders, testing the tightness in his neck, then raised his hands over his head, extending his rib cage. "Ow!"

"Take it easy."

"You gotta admit, it was a lot of fun." His grin was infectious, and she found her mood lightening.

"What was it you called me when the guard showed up? Ellie Mae?"

"I think it was Mary Ellen, but don't hold me to it."

They shared the moment — the adrenaline; the all too easy heart-pounding escape in the middle of the night; firing the van before nudging it into a sinkhole; the short "get to know your new guy" visit with her mother. And finally, the drive through dawn and daylight to a mid-range motel whose minimum requirements were cold beer and a pool. The perfect picture of a couple taking a weekend vacation. The daily news had shifted back to pointless political commentary in less than twenty-four hours. Videos of billowing flame and terrorism-speculating talking heads were replaced by tales of corruption in the Senate and different talking heads.

"Well, while the world was looking for someone named Mary Ellen, I was scared shitless we'd be hunted down a la Bonnie and Clyde."

"Too close to the edge?" He didn't get an immediate answer. "You know, Grieg was great in Latvia. You can maintain your CEO role. He and I or he and Chu can—"

"—can get yourselves thrown into a Third World jail or shot to death or both. And for what? Florida Gas Transmission had both pipelines repaired in a few weeks." She set her jaw, jutting it slightly in grim determination. Any of her former litigants would have been reminded of the old Parker in the moment before getting into the face of a deponent, going for the jugular at the end of the feint and parry questioning; going for the big ask under oath. "No, we need to move on."

"Parker. Stop. You're getting worked up. Our next gig is a piece of cake."

"So what? You drop a powerline serving the Turkey Point nuke, and then what?"

"What? We bring attention to the fact that the plant is leaking like Fukushima at high tide!"

"OK, suppose Florida Power shuts Turkey Point down, and they buy more gas. Kind of negates our demonstration of Florida's dependency on natural gas doesn't it?"

"At least they shut the goddamned thing down."

Through thin lips she hissed, "We can do that with peaceful protest. With scientific disclosure on the evening news, the *Miami Herald,* and *The New York Freaking Times*! Shame the bastards into taking it offline, or at the very least fixing it." She continued mumbling sub-audibly about officially acceptable levels of atmospheric leaks.

"Honey? What the hell? What's wrong?" She had rarely exposed her passionate investment. Her normal mode was calm analytical study, plan making, execution, and, as often as possible, shared joy in accomplishment.

She turned to look back out at the ocean. Her chest suddenly swelled with emotion, hitching her breath. She found herself clamping her jaws hard to prevent speech.

He continued. "Where's this coming from?"

She turned back to him, water rimming her eyes. "I don't want to lose this!" She swallowed a sob. "I don't want to lose you! Or Room for Tomorrow, or..." She hesitated. "Maybe the Verts.

They may still have a place, but please, not as environmental terrorists."

He dropped his head and away, a picture of disappointment. "Well, what then?"

After a too long silence, she gathered her composure and responded. "The Verts can publicly renounce the violence that made them a household word. Hated in too many households. We, you, can disappear all digital records of their existence." She frowned again; her forehead furrowed. "I wanted to do more than squirt water at whale ships. But technically, what we've been doing in the VSF is the same sort of theatrics."

He raised his head again, meeting her fixed gaze. "Only more dangerous." He held the eye contact. "Let's think on it. OK?"

"Who besides Brayton, Grieg, and Chu know who the Verts are?"

"Meg is about it. My only field operatives with more than burner phone connections won't be born for over 150 years." His gaze became unfocused as he mulled over the list of contacts and their modes of contact. "Yeah, just Meg and of course, Brayton." He turned to her; his look intense now. "Park, we're going to need to step up electronic countermeasures." She nodded, understanding. "I'll get Brayton to scan and clear the sixth-floor inner conference room. No windows, no electronics of any kind. We're going to need better anti-surveillance measures than weekly bug scans, jammers, and talking over the surf."

She forced her thin, grim, determined mouth into a semblance of a grin. "We'll do what we have to do." She reached out to him, to caress that leg again. She felt his calf tighten then relax under her fingertips. Her mind cleared, headache gone. Perhaps it had been from worry, not wine. "Carl?"

His mind was still on security. "Yeah, we can never let our guard drop."

She lost herself in the endless line of incoming swells. Their smooth surfaces reminded her of the rippled, clear plastic sheets her dad had installed on his on-the-cheap greenhouse in the back yard. He had been the early influence, the first real lessons in solar heating for free or cheap. Composting in a county that hadn't yet begun to recycle. The train of thought brought her to their next solar project and to trains. They were to meet with Euro-electric execs in a sun-favored location on the Adriatic. Long ranges of too-steep-to-farm low mountains faced south-southwest, perfect for large arrays. "Carl, I've been thinking about the trip to Croatia."

"Yeah, too bad the Orient Express doesn't link up; that would be, well, romantic as hell."

"Romance isn't what I was thinking of." She let a pause linger. "I want to meet with Jozef Jaager next week in Amsterdam. He responded to my email from a few weeks ago. He's retired now but still has access to the surge gates. He's a mini-celebrity there. He said he could clear me for a close-up tour and even line up some money partners who had helped the Netherlands ministry finance the project."

"Wow, that's pretty cool, babe." He let the implications sink in. "Sure, you go to Amsterdam, and I'll go to Croatia and meet, uhm, what's her name?"

"Anna. No, Anika Kovačić." She poked at the firm flesh of his calf. "Behave yourself, I heard she's a fox."

"I heard she's married to the top guy at the Croatian State Department." He retorted.

They exchanged looks that spoke of an earlier temptation, then resumed their silent study of the incoming swells. OK, she thought, we aren't married. But dammit. Focus now on what's important. Her next task was to try to save Bangladesh from itself with engineering solutions from the Netherlands; Carl's was to bring renewable energy to Eastern Europe.

"Carl?" Her tone shifted, less business, more relaxed.

"Yeah, babe. What is it?"

"If you'll refill this wine glass, I can get more than your calf muscles to relax."

~ ~ ~

Odie Peck pulled off his earphones with a crooked twist in his mustachioed grin. He looked around the night-darkened walls of his Western Kentucky home and compared his humble surroundings to the plush California cliff house on his monitors. He took a swig on his longneck and muttered to himself. "Sons a bitches gonna see that shit in the surf." He looked at the I-beam struts supporting the deck and pool beyond the edge of the cliffside house and planned how to access them. He made mental note to get the HMX and Claymores from his stash before driving back to LA. Peck's mouth twisted in malicious intent as he stared at the house. *I'd like to massage her calves.*

He appreciated the technology he'd installed on the rental house leased near Parker Parrish's cliff house. Its recording and interactive telephonic abilities were world-class. He popped the top on the evening's second longneck and took a long, cool pull. Setting it down, he waited for the belch and let it go. There was no one around, no reason to be polite. He'd been alone now for more than five years and accepted that the travel demands of his new employment with the Abadons would have tested the limits of most relationships.

He backed the video file up to the scene when they'd come out onto the patio. That Brayton bitch had found almost all of the mikes in the house, and the remaining one in the foyer was too distorted by echo to be useful. But the audio dish and camera aimed at the patio had done their job. Half a billion bucks, eh? He whistled out loud. "Jesus in heaven, that's a crapload of cash."

The sound of the whistle seemed to be the breakpoint he needed to spring into action. He got up and moved to his personal computer on the kitchen bar. In a few minutes of keystrokes, he had tickets for Grand Cayman. Perhaps the Abadons would believe him when he showed them deposits that matched their disappearing funds, amount for amount and date for date.

He wasn't sure how much he'd need to bribe, swindle, or threaten a still unidentified source, but he was going to enjoy the trip. There was always someone with a weakness for drugs or women. He smiled again and finished off the longneck. And there were some real lovely cinnamon-skinned honeys down there.

~ ~ ~

Parker was finally beginning to relax. It might have been the ever-present vibrations of the Boeing 747-400 or the second glass of brandy, but the semi-reclined first-class sleeper seat was beginning to feel more than comfortable. Only recently accustomed to first-class travel, she had thought she would never be able to sleep on an airplane. But this was comfort. Her late booking had forced her to take one of the double berths usually assigned to couples flying together, but the well-dressed man in the adjoining seat kept to himself and an iPad full of financial graphs. She pulled out the removable screen from her mini laptop and began scanning information on the Dutch flood control gates.

Her info tour scheduled for the following day was with one of the principal civil engineers, now retired, who had designed the award-winning marvels that kept the sea out of the low country. Named the Maeslantkering, or Maeslant storm barrier, the swinging door floodgates protected Rotterdam's huge port investments and thousands of hectares of prime farmland. She understood that structures like these were quite expensive, and the ports of New York, Jacksonville, San Diego, or Baltimore would make those investments. But the rest of the world would be in need of new locations or lose their ability to withstand catastrophic weather entirely. Problem ports such as Miami or Chandpur might have to be abandoned. Chandpur was the heaviest on her mind and her current distraction; Miami could take care of itself.

"Do you mind telling me where that delta is?" The businessman to her left was looking at her laptop screen. He'd been taking side glances for a while, but this was his first overture.

"Uhm, sure. It's the Ganges delta, Bangladesh." She felt surprised and awkward with the question coming before introductions. She was about to explain why when he spoke again.

"I thought so, but I wasn't sure if you'd had the map oriented north-up. It's known locally as—"

"The Padma, thanks." Irritated at his intrusion, she checked the north arrow in the map app's info box and righted the tablet slightly. "I think that's about it. You know the area?"

He set down his tablet, lowered the lounger's leg rest, and adjusted himself in his seat to face her better. "I'm with the World Bank. We've done some work in Bangladesh. They're a complicated lot, and their problems go all the way back to the partition," he paused, "and before that."

She thought of the powder keg that would become the flash point of the last great war. She was nodding in thoughtful agreement when he added, "They got a raw deal, the Muslims. They got desert and swamp. Sure, there's some good land in the mix, but most of Afghanistan is a mountainous desert wasteland. On the other side of India, Bangladesh is dominated by either desert mountains or salt-prone lowlands. Its coastline is the world's largest mangrove swamp, and that is shrinking yearly.

One good bullseye hit by a typhoon and millions that live on the mangrove region's margins will drown."

Parker was well aware of the extent of the coastal marshes and complicated river outlets that comprised the delta system at the mouth of the Ganges. Her own team of out-of-the-box design engineers had come up with a plan to construct levees across the lowlands of Bangladesh. The full installation would cost billions that neither Bangladesh nor the World Bank could come up with. She also knew that the levees and structures that the World Bank would build in the 2070s would be insufficient. But her corporate background contacts had come up with a business model based on co-generation of solar and sustainable farming that might preserve the region even after a four to five-foot sea-level rise. The plan would allow farming as long as it was possible on the seaward side while transforming the often inundated lands northwest of the proposed dikes.

She chewed her lip briefly and then turned in her own seat to face him. "I have a team of engineers, economists, and other peripherals working on just that problem." She appraised him further — exquisitely cut three-piece suit, loosened school tie with gold on navy embroidery, and a meticulously trimmed mustache. His light gray eyes narrowed slightly as he matched her searching eyes. But when a benevolent dimple appeared on his right cheek. She said, "You're World Bank?"

"Yes, forgive the intrusion, I'm Oscar, Oscar Morand." He extended a hand as if to take her fingers in the European fashion leading to a peck on the knuckles. She instead inserted her hand in his and shook it firmly.

"Parker, Parker Parrish. I'm with Room for Tomorrow. You may have heard of our international fund raising. We —"

"Yes, Yes, I certainly have heard of your efforts. The Jutland wind farm project raised many eyebrows at the Bank. And the scope of the Nebraska project you're planning, well —" His shrug was pure Parisian. His accent was an undefined euro base with inflections from both German and French.

She tried to hide budding excitement. Chance meetings like this could do so much more than formal spreadsheet-and-prospectus-laden board room briefings. Behind a thoughtful smile, she was formulating an appeal when her phone chimed, signaling an inbound email. She blinked at the interruption but continued,

"Mr. Morand, may I call you Oscar?" She acted on an answering smile and continued, "Oscar, we're looking several decades into the future at a refugee crisis that could have world-shattering implications." Her mind had been replaying Chu's doomsday chronology. "We think the Ganges delta is important not only for fisheries nurseries in the mangrove wetlands but also for the arable lands further inland that provide food and employment for hundreds of thousands—" The phone chimed again, this time as an incoming message. "Excuse me, Oscar."

She swiveled around enough to grasp the phone and activated the screen. In two taps she opened her messenger. Reading the message, a cold shiver ran the length of her spine. "Kee-rist!"

Her exclamation was on the edge of sub-audible, but Morand picked it up. "Something wrong?"

Parker moved her lips, but nothing came out. Beads of cool sweat formed at her temples. Her breath threatened to stop entirely. "There's been a..." she halted, disbelieving. "A good friend of mine," she covered her mouth with her free hand, a hedge against losing control. "A partner actually, has been involved in a horrible accident."

"I'm so sorry, Miss Parrish." Parker looked up at the comment to see Morand's face awash in genuine concern. "Do you know if he's all right?" Then after a moment, he added, "or she."

"I don't know." She looked back down at Megumi Sakoda's too brief message and read the two terse lines:

[Parker, train derailment Triest to Rijeka,
23 dead, Carl missing].

Barely audible, scream suppressed, she heard her disembodied voice say, "I should have been on that train."

~ ~ ~

When the woman excused herself to get privacy in the plane's upper deck lavatory, Morand fished his own cell phone from his vest pocket. He dialed up KLM's Gogo service and tapped his own message out.

[Hon. VP Koolapoor, at your convenience, need to talk about new project, Bangladesh].

Upon hitting send, the simple message sent to the Vice President for East Asia and the Pacific of the World Bank had as much potential as the first domino in a chain, or the latch on a garden gate.

Morand settled back into his seat, grasped his tablet, and called up a view of the Ganges Delta. His grin transformed his face. He loved dealmaking.

31

A resounding metallic crash in the distance woke Carl from a deep protective layer of subconsciousness. Awareness slowly resolved as a thickly flowing fog thinned, allowing thought. Pain, oh Christ! Pain emanated from everywhere. Carl relaxed muscles that wanted to contract against intense trauma. Focus! He felt his skull reverberate with the rhythmic pounding ta-dump of his heartbeat.

He tried to raise his head and immediately felt resistance. Exploring unexpected opposition, he flexed an elbow and again had the impression that he was bound but not tied. Methodically he moved arms and legs outward. An immediate sense of confinement caused an involuntary surge in every muscle group that resulted in increased levels of general and specific pain. Not tied but confined? *I can smell the stench of my breath that smells like, what? Blood? What the…?*

He relaxed and attempted to take inventory as his faculties came online. *OK, I can't see, and my breath blows back in my face. I can't move anything more than inches. OK, slide right arm from under back. Good, it's not bound!* Muscles and tissue complained everywhere. Straining to move, he realized he was wrapped in something dense and only slightly yielding and crinkled like vinyl.

He flexed at the knees. The motion sent a draft of air from the depths of hell. *God!* His nose took in an incredible stench: panic, sweat, urine? *Have I peed myself?* Above the smell of fear, understanding emerged from panic. *Oh Christ! Is this a body bag?*

Pulse and breathing quickened, the rapid susurrations of his breathing were amplified by the confines of the bag. *Chill, you're getting lightheaded, running out of oxygen in this bag.* Survival skills from his years on the streets of East LA cranked in. *Calm down. What's that newly ubiquitous phrase? 'Keep calm and carry on?' Great! Churchill never woke up in a body bag.*

He attempted to roll sideways to the left, but something soft blocked the motion. Experimenting again, he moved his right hand slowly to his chest. He groaned as a wracking spasm spread outward from some bodily insult in his lower abdomen. Catching his breath, he slid his hand inside the scant spare room upwards to his face. Stubble? He stroked his chin again, reaffirming. *I have stubble?* One- or two-day's worth. Turning his fingertips upward, he felt the inside crenellations of a zipper and followed the line. Against complaints from his ribcage, he turned his head to make room for his hand to explore the upper reaches of the zipper. He could sense the air in the bag getting more foul with every exertion and exhalation. His heart rate had escalated from near coma to just below panic, and his breathing was almost out of control: he had to get the zipper open. He found the end of the zipper track almost fully closed and hooked a finger on the inside of the pull.

He paused, allowing paranoid reconsideration. *Why am I in this bag? Who put me in it? They must think I'm dead. Could whoever packed me in here, be watching? Would it be good to be alive if dead was the intent?*

Carl's aching body stretched involuntarily against the confining plastic and settled back onto hard metal. *No motion? I'm definitely not on a moving train. But whiskers! I have whiskers now. That means at least two days.* Noises take on pattern. Listen! Low voices or a radio or TV somewhere distant. Certainly not English. Listening more intently; not French or Spanish.

He took stock again. *I'm clothed, feels like jeans and a T-shirt. No shoes. But string on a toe. A toe tag? Someone thinks I'm dead. I am NOT dead!* He tilted his head back and felt a thin puff of cooler air. A thin wedge of not quite so black appeared at the limit of his vision as he craned his neck backwards. *If I can see that hole, it's not totally dark wherever I am. Softness to my left. Another body?* He reached for the opening as the muscles of his right arm argued against every exertion. He hooked a fingertip into the opening and tried to pull the slider down, the bag wrinkled but did not release.

Carl pressed his head into the crown of the bag and pulled again. This time the slider slipped down a few teeth. He smiled stupidly at the small victory. Small victory, he thought, because if he couldn't get some fresh air soon, he'd pass out. Too much flex in the one-size fits all bag. Straining with extended feet and neck to straighten the zipper track, he worked the zipper slide back and forth. The zipper pull was on the other side of the track, the side that alive people use. It wasn't intended to be pulled from the inside. It slid two or three incredible noisy inches. But there was air. It was fetid and only a little more oxygenated than the interior of the body bag! Something worse than his fear and stench filled this place. Rot and corruption. But mixed with the foul odor there was oxygen.

Against gagging olfactory overload, he filled his lungs, then swallowed a spasm to vomit. "Uhh," he panted, surprised almost at the sound of his voice. Still not sure if he was here by mistake or intent. The recent Paris bombing resurfaced as much less of a coincidence. They said that if you were hit twice by lightning, it was not coincidence—God intended to kill you.

Carl exhaled thoroughly in a deep exhalation and inhalation, fighting messages from his diaphragm to heave in disgust. The thick veil of confusion began to lift. Air, far from sweet, but air. A glimpse of memory flashed; visions of darkened streets cobbled in granite. Europe, dear troubled Europe. He stretched out again and pulled the slider another few inches, clearing his face. The sound of his panting breath reverberated in a closed space. In a near dark that was only a little lighter than the depths of despair, he scanned the enclosure. Dark, featureless walls in a long narrow room.

Before he pulled further on the zipper, caution prevailed. He needed to stretch a still soupy intellect to think of the last place he could remember. France? Paris? No, the train from Florence to Trieste. *Did I get through the border crossing? To...don't remember. Yes, transferred to the train to Rijeka. Conference Tuesday with the Croatian finance and energy ministers. Don't guess I made it!*

Have I been disposed of? Tossed out? He felt a wall of some kind inches away from his head. He pulled his right arm free of the zipper and felt cold metal. Slapping it, the reverberating bass thump was much louder than he'd expected. A metal wall. A shipping container? He hit it harder with the meat of his palm. Another bass boom, louder than the first. *Should I be announcing that I'm alive or begging for help? I still haven't figured out the who, what, why, where, when. Hell of a reporter I'd make!*

With the other arm out, he reached out to the left. The soft lump he'd bumped into earlier resolved to his touch to be another body bag. "Oh, shit!" His mind began to race, punctuated by, "Shit, shit, shit, SHIT." Not alone, and not in good company. Turning toward the adjacent orange plastic lump, his questions flowed unimpeded by answers. Except one: no, I do not want to unzip that other bag. The air in here is putrid enough without opening that bag.

Straining against his complaining body's inclination to not move, Carl moved. Some trauma, cracked rib? Torn ab? He sat up to spite the pain. Lights flashed in his retinas as intense abdominal fire flared. He paused to let his breath slow down against the waves of agony. He unzipped further, down to his ankles, took a deep breath against the expected spasm of complaint, and rolled to his knees. Before the screams from damaged tissue could override willpower, he slipped out of the bag onto all fours.

He felt and looked like a beaten dog. Breathing rough and irregular, bile coated the back of his tongue. An urge to heave against the foul air had to be suppressed. *It will hurt more than it's worth.* He realized that his eyes were closed tight in an effort to concentrate on what hurt most. *This nightmare needs to end.* He opened them and looked up. There were more orange lumps scattered; no, arranged in poorly aligned rows. Twenty-three, twenty-four, five, six. He counted twenty-seven. *What killed or almost killed twenty-seven of us? Who is us? Are we all dead?*

"Hey!" He called out a tentative hello to the twenty-seven dark orange lumps. "Hey!" A little louder. None of the bags twitched. *I've got to get some fresh air.* A tiny pin prick of light was the best chance for outside air. Getting to it would mean having to move or climb on body bags. He scanned for alternatives and didn't see any. "Well, fuck me." He crawled toward the point of light over softly yielding forms zippered into orange baggies. The faint light projected a dim scene on the opposite wall.

Staring, it seems ridiculously skewed — no, upside down. *Of course, it's a camera obscura.* The dim scene on the opposite wall was outside, projected upside down and reversed on the opposite side. Inclining his head down helped make sense of the scene. A street scene, parking lot, vehicles, parking spaces. Carl remembered the need for fresh air. He kissed the wall, placing his mouth over the small hole and inhaled. His lips felt a small raised edge of torn metal. He realized it was probably a bullet hole but didn't care. Why wouldn't there still be spare bullet holes in Croatia? Not all signs of war disappear with spackling and paint.

Oh, sweet air! Breathing in through the hole was not tenable for long. He took several more lungfuls anyway, knowing that he was kneeing some poor dead bastard in the neck to lean here. He paused for a moment of concentrated thought, tensed, and stood as fluidly and quickly as possible. The sudden move nearly made him pass out. As he stood, unsteadily wavering, he shuddered at the thought of a collapse among the corpses. Hand against the wall, he inventoried the many levels of pain signals. The worst of it was coming from his right lower rib cage, but there were sore spots almost everywhere. Exploring, he found what he suspected were two cracked ribs and a wound. A hole! Between those broken ribs was caked, dried blood. His fingertip came up slick with fresh blood, black in the darkness.

My blood! His knees weakened, and he fell toward the wall. The thump of impact was deadened by a reinforced structural rib, but the wall yielded slightly. He fell across two of the corpses, but his sense of the metal overcame a stabbing pain in his lower ribs. *The wall is not steel, it's aluminum! Not a shipping container… it's a semi-trailer. Hope springs eternal! Some trailers have access doors. If it's a reefer, there's hope. It's dark in here — either end could be the back of the trailer.* He picked one, moved toward the end with the most body bags, concluding that they wouldn't have carried mass casualties any further than necessary into the trailer.

Stepping over bags, supporting his weight on the wall, he made it to the end of the trailer. It is the back! As he approached the back wall, he saw a thin line of light at the base of the doors. Light-headed still, he leaned into the corner of wall and rear door.

That sound from earlier, it's more distinct here. *It's a broadcast of some sort. Talking heads.* The news sounds like the news in any language, a mixture of bored and emphatic, but not excited enough to be sports. He knew he didn't know the language. He wondered aloud, "Am I guarded? No, they think I'm dead. All twenty-seven, uhh, twenty-eight of us." He turned to peer down the barely-lit length of the temporary morgue.

Why is this trailer not refrigerated? Why were twenty-eight supposedly dead — *Footsteps! Should I call out? Bang on the walls?* "God, what am I, accidental death or targeted homicide? Paris, now this?" The footsteps tapered off into the white noise of tinnitus.

He felt toward the center of the door for the seam. There was none. He stroked the door's surface further, finding multiple seams, only horizontal and a few inches apart. *It's a rolling overhead door.* He leaned, pressed against the door, and pushed up. Mechanical noises complained of poorly oiled ball bearings, and a thin band of daylight revealed his feet with a rectangular tag on the right big toe.

The mysteries kept deepening, but there was fresh air to be had. Dropping to his knees, he remembered how many places hurt. Nearing the slight opening in the rolling door, there were increased sounds — distant traffic, mechanical sounds. The soundtrack, still less than fully audible, was definitely not English.

Lying on his left side, mouth to the thin crack at the bottom of the rolling door, he assessed his situation. "OK, I am presumed dead. Blown up and presumed dead." With fresh air clearing his lungs and his head. The rhythmic tapata-tapata of rails joined the pounding of blood in his temples. He realized the ringing in his ears was too loud for tinnitus. He remembered a terrific jolt in the carriage, and then the crescendo roar of a train in moving immolation as his car passed through a fireball. He was thrown violently, helpless as a tossed limp doll. Lights sliding beyond the windows were moving in the wrong directions. The carriage must have been climbing onto its nose. Screams, tearing metal. The car itself abruptly thrown to the side. He remembered grasping at any fixed object as he slid ridiculously across a row of luggage racks and fell onto the ceiling. He had come to a stop with the edge of a light panel tearing into his torso at the lower line of ribs. Someone's inert body landed on his legs. Then nothing until the wake-up call in the orange body bag. His finger again traced the broken ribs and found the hole in his shirt. More fresh blood.

Carl pushed against the rolling door. It moved grudgingly, but it moved. He lowered his face to the floor and wedged arm and elbow into the scant opening. Push! The door rose a few more inches, permitting him to look out. A streetscape in dressed in gray overcast. Europe. Parked cars, ambulances with green crosses. Ambulances? Hospital or morgue? Footsteps! "Hey!" He gasped and, sensing that it was only a gasp, tried to yell. "Hello!" He took another deep breath against the pain of yelling and steadied before his next try. The footsteps were running now, approaching. The face was young, clean, and worried. A nurse in scrubs. Her eyes widened in shock. *I must look like hell.* "Ayuda me!" Then, "Help me!"

Mercifully, he blacked out.

32

Parker turned from the nurse's station as the only person on staff with rudimentary English pointed down the hall to Carl's room. She restrained the urge to run to Carl's room but managed a fast walk that didn't reveal her hyper anxiety. The sight of a near-total body wrap of burn dressings forced an audible gasp as she reached the doorway. The unfortunate figure was a caricature of the one-leg-suspended near mummy.

"Over here."

She turned and saw Carl's smiling face. His grin seemed forced and soon fell away as he tried to beckon her over to his bedside in obvious discomfort. He was sitting up with the aid of an inclined mattress. A large bruise discolored his left temple, and his waving right arm trailed thin blue tubing.

"Come here!" he insisted. "He won't mind." He gave an impish nod at the mummified patient in the other bed.

She let a sigh carry away most of the hours of dread. Language barriers and her inexperience with European telephone tech had prevented her from getting a call through. Without a translator handy on short notice, she felt ill-equipped to order Coke much less name a barely pronounceable city's hospital. Now seeing him supported by the folding bed's mattress, he looked far better than she'd feared. "Damn you! You scared the crap outta me!"

"Scared the crap out of myself." His attempt at a small laugh forced a grimace as a wave of pain spread outward from the patched flesh over his lower right ribs. "Woke up in a body bag."

"Body bag? They thought you were—"

"Dead. Apparently, I'm hard to kill."

"Who do you think? Russian mafia? Triads? Who?"

She had been thinking the worst for the last two days. Paris could have been an accident; wrong place, wrong time. But twice?

He only shrugged. He appeared to be thinking of a response, but it didn't come.

"Carl! Come on. Anyone who might have hacked our travel plans up to a few days ago would have thought we were both going to be on that train." She was convinced; no longer speculating on the immediate danger she felt manifest. "Who else do you think might be trying to take down Room for Tomorrow?"

"Oh, I don't know, any of a half dozen or more major corporate interests from PeMex to OPEC, and yes the Russians, or their syndicates." He almost shrugged from habit but caught himself. "I guess that's the same thing. Anyone with a vested interest in making money on sucking the last of the world's fossil fuel from the ground."

"What's a peemex again?"

"Petro Mexico. But really, any state-owned or corporate interests out there are going to want us to fail. And then there's the Abadons, or the Munsons. Remember the Paris explosives were linked to Kentucky. Odie Peck was on my trail, and he's from Kentucky."

"But to resort to murder? Not just murder but terrorism masking murder. They said there were eighty-eight dead with more in critical condition. That's not murder; that amounts to corporate-sponsored terrorism." She felt herself yielding to a manic edge that she almost never allowed to show. She had allowed mania to take over a few times in her life: the weekend prior to the bar exams, the last fight with her second husband prior to the divorce settlement, her brief foray out the back door of CalStation, and on the flight to Brussels when she was absolutely powerless to speed the plane on to its scheduled landing. What he had just said finally registered through her buzzing thoughts. "Who is Odie Peck?"

"Hon, there's a lead. Almost certainly confirmed. I tracked an inside security presence in the Abadons' Cayman accounts. The one the Tax Man doesn't know about. He was tracking me. His name is Odie Peck, from Kentucky. I mentioned him a while ago; he was snooping in the same accounts I was in."

"You found him in there? In those same accounts you've been hitting?"

"He found some of my tracks, but—" He saw the look of alarm spreading upward in her eyebrows. "But he didn't see me!

Calm, calm." He made downward suppressing gestures with his left arm. "I sent in a worm to paralyze his ability to follow me, followed by a false set of tracks that will send him to a fake account that is hosted by the superior Hong Kong Triad." He tried a smirk, but it hurt. "He should be looking for good Russian and Chinese translators to work with."

She could tell he was just trying to be reassuring, but she'd seen Brayton's report. The work he had sent to Brayton for analysis proved that Carl's presence had been backtracked to Southern California servers, possibly to Carl's machine. Validation of Carl's theft would be cause enough for murder.

She gingerly took his hand. "Hon, Brayton said the worm didn't work, entirely. It's good you know who you're up against, but they are getting close, very close. We have to back off our Verts Sans Frontiers. Close it down."

"So…if it is Peck, what if he figured out where we got our start-up funding?" Carl closed his eyes as another wave of pain stiffened his torso.

She watched the pain tighten his face. He clearly wasn't in condition for these conversations just yet. "Oh Jesus! I'm sorry. I come in and lay that crap on you." She sat on the edge of the bed. "Is this OK?" She leaned forward to smooth his hair. He pulled away as her hand got too close to his forehead. "Oh, sorry."

"That's OK. Just don't touch." A wry smile grew again, irrepressible. "Almost everything hurts, Park. Hell, they thought I was dead. You said there were eighty-eight dead reported?"

"Yes, but I guess there's one less. You made it."

He motioned to the near mummy across the room. "There are some more down the hall who are in pretty bad shape. Some of those might not make it. One guy had his leg sheared off above the knee." He closed his eyes, his pupils racing and erratic under the lids. "If," he let a rib-stiffening wave pass, "If I hadn't gone for a drink in the dining car, I would have had it." His eyelids squeezed tight, perhaps against the vision. "I would have been blown apart and immolated." He seemed to sink further into his pillow. She waited, hand on his good arm.

"I have some bad news."

He looked up at her. Exhaustion-darkened sockets framed water-rimmed eyes. "Give it to me."

"Amrat is among the body count. He didn't make it."

"Dammit! He was excited about the… he was…" He stared at the ceiling. In the US, there'd be foam tiles there. Above him, a patched and peeling ceiling evidencing numerous paint-overs. "What about—"

"Meg has already contacted his family. She's also gone to Florida State to talk to Amratiputhi's energy sciences department head and his lead researchers about continuing with the hot salt storage project." She gently increased pressure on his forearm as another pulse of pain forced his eyes closed. He shouldn't be up. She considered a kiss on the forehead before leaving, but his first comments came back to her. "You said you woke up in a body bag?"

"Yeah, pretty scary shit." He paused, blinking. "That was a very bad time."

"WTF. How did that happen?"

"Hell if I know. I just know there was a trailer full of us who were in line for our turn at the morgue. How many places have room for eighty people at their morgue?"

"If anyone does, you'd think Croatia would be in that club."

"Seriously." He coughed, winced, tried to joke through the pain. "But I thought they favored mass burials."

She wasn't receptive to humor and got back to the reason for the trip. "Guessing you haven't been able to catch up with Anika Kovačić?"

"Well, I did miss our first appointment. The nurse said I could call later. The one that speaks English."

"Sort of speaks English?"

"Yeah, that one. Actually, a few of them try a little here and there, I don't have any complaints. They've been good to me."

She smiled. "Doesn't hurt that you're handsome and rich." He looks like hell, she thought. "Should I leave you alone to get some rest?"

"I could sleep. I was almost going to dose myself and go back under when I heard your voice down the hall." He lifted the small wired button device that would allow him to up his pain meds on a demand basis. She watched his thumb depress the button. After a brief pause, he tapped it again.

She knew she didn't have long. "Tell you what, I've rescheduled Brussels for next week. I'll meet with Anika Kovačić to get the proper contacts for a local steering committee to get that solar farm going." She glanced at the sunlit scene outside. "When you're feeling better, we can come back and talk about the hot salt storage installation. You understand that tech better than I do." She turned back to Carl and noted his eyelids at half-mast. With more hope than she'd have anticipated a half hour earlier, she leaned over and kissed him on the unbruised side of his forehead. She was going to suggest they ask the board member charged with marshalling energy storage efforts, to come to sit in as well, when she saw that his lids had closed.

She noted that scars from the Paris glass shards still remained as thin red lines. By the time she had straightened and stood, he was fully relaxed in an oxy-dream state. She went to the window and adjusted the slats to closed and dropped them to the sill. At the doorway, she stopped and turned back. "I love you, you live-dangerously, full-speed-ahead, SOB."

~ ~ ~

From his seat in the Abadon brothers' waiting room, Odie Peck listened to the sibling warfare going on behind closed doors. He knew that whatever the outcome, he was next. One of the two would be calling him into his inner sanctum.

The two brothers had long-ago split on their support for his work. Andrew had no problems with his technical counter measures, the listening devices and the counter attempts at hacking into RFT's donor lists. After all, Peck had been somewhat successful at getting access to Room for Tomorrow's server bank, but it had taken months to establish even a tenuous cash flow connection between the unresolved accounting deficits in overseas transfers and the sporadic increases in the coffers at Room for Tomorrow. Andrew had also been unimpressed at Peck's inability to find a way to get an understanding of the mysterious activities of Verts San Frontiers. Peck had insisted the two were linked if for no other reason than both organizations appeared at the same time and seemed to have phenomenal funding and organization. Andrew Abadon was specifically not into authorizing murder and mayhem.

Sitting there, waiting on the outcome of the argument and their decision on his next target, he scanned down his list of latest achievements. He'd found a way into the RTF server to download and decode contact phone and email lists, donor lists, and their records. He knew who the major repeating large-amount donors were, and he had developed a social profile of the twenty-five to hundred-dollar donors. There was little he could do with this data, but he could compare donation flow to the meteoric rise of Room for Tomorrow, LLC. He felt he had demonstrated a connection between cash flow and the Cayman cash account errors. The sound of argument subsided, and he found himself leaning toward the double mahogany doors. He thought, *Come on Richard. Plead my case, you know you want me to finish this.*

With a soft click, the handle dipped. It didn't open immediately, and he heard. "Yes sir, he's waiting just outside." Peck stood up in anticipation. He had finally noted the nameplate on her desk, Margaret Campbell. The shapely silver-haired guardian of the gate stepped halfway through the doorway.

"Mr. Peck. Please follow me." Her tone, less friendly this time, did not invite conversation.

He did, following her to the right, past an empty conference table, toward the ten-panel mahogany doors leading to Richard Abadon's office. *Yes! I did it.*

Richard Abadon, the younger and usually more aggressive partner, was standing and leaning on the front of his desk. As Peck entered, Richard pulled a folded paper from his jacket's vest pocket.

Peck noted that Richard didn't break eye contact once as he unfolded the several stapled pages and flipped to the second page. He scanned only a few scant words and recited from memory. "First, we conclude that the train bombing Tuesday last was not the work of radical Islamist groups. Despite their rush to approve the act and to accept the responsibility as a result of their lone wolf media campaign, we conclude the train bombing was targeted against a passenger." He looked back down to the page for detail. "Further, our sources inside Interpol, who are assisting Croatian authorities, have traced explosives residue to HMX purchased by a French contracting company in Porto di Gioia Tauro, Italy. Same company, Ursa Demolizione, reported theft of identical batch of HMX to local security offices, per the head of the Carabinieri."

Richard glanced down at the intel report and read in slow, clear, and painfully concise tones: "Lastly, based on similarities in recovered internet communications, we strongly suspect, and are working to confirm, ties to the Rue de Rivoli bombing in Paris last month. Confirmation of this link will almost certainly be made by Interpol."

Peck felt uncharacteristic sweat forming in his armpits, even his temples. *This can't be! Those were purchased locally.* "What does an Italian mining company have to do with…?" *Goddamn Tarique!* Peck was confused.

"Really doesn't make a helluva lot of difference at this stage, Mr. Peck." Richard Abadon folded the second sheet over and read items from the last page. "The complete list of identified survivors follows and does not include subjects Parrish or Reyes. Inquiries at the hospital in Rijeka. Subject Reyes admitted with non-lethal injuries. Local celebrity attached to his person as the initial triage concluded dead at the scene. Subject was bagged and stored for processing at the morgue."

"But my report on that list—"

"Was wrong." Richard's voice carried conviction.

Peck, knees weakening, leaned on a brass-studded red leather chair. Then as the weight of the news settled, he sat heavily on the chair arm. "It can't be. The charge was in a satchel beneath their—"

"Stop!" Abadon's command voice was intimidating and successful. "Your other objective, if my information is correct, was not even in the vicinity. She was, in fact, on a trans-Atlantic flight to Brussels." Peck settled fully into the seat of the chair, leaning forward, elbows on knees. He looked up, ashen, mouth open about to speak. Abadon continued, a finger to his lips. "Don't ever vocalize specifics in this office or this building. You know the rules, you idiot! We discuss contracts, terms, oblique methodology, but never project specifics. Is that understood?"

"Yes Sir." Peck's weak reply was barely audible. He tried again louder. "Yes, I understand. Operator error, sir. Instructions were precise, the uhm, operator clearly repeated back instructions. I had no doubt that—"

"Well, your operator seems to have had some doubt."

"Yes, I— Where does that leave us with respect to the project?"

Abadon flattened the three pages, plucked the staple with practiced fingernails, and slid the stack into the slotted throat of his desk's built-in shredder. Neither spoke for the few seconds it took for the mechanical teeth to shred the privately obtained report on the now world-famous bombing of the Zagreb to Rijeka night train. The explosives had detonated as the train entered a narrow defile; most of the carnage was caused by the telescoping cars piling onto others. The bomb, considered to be amateurish, had blown out the walls of the targeted car and dislodged the car from its wheel truck. The news cycle had been limited to floodlit night shots from a lone helicopter piloted by a tourist agency despite the government's insistence that it stop streaming news video. The low-quality images were more than enough to tell the world of the horrific crash site and make the chopper pilot and cameraman temporary celebrities.

"The project," Richard answered in slow even tones, "your project, Weed Cutter is canceled. All records are to be purged, shredded, erased. I would personally appreciate it if any hard drives you may have be destroyed. Expense account it if you must. I believe you've been adequately compensated for any perceived losses on that account."

Peck couldn't believe it. That bastard Reyes is alive and now the possible coincidence is going to lead back to Tarique. That Arab shit will most certainly be willing to sell me out for a reduced sentence. Choices? Kill Tarique Al Khayr, before Interpol finds him? Retire to Guatemala? No, too hot. Argentina! Fuck, and now heartburn.

"Peck, are you listening?" He hadn't been. "Odie, snap out of it. And if you would, please hand over that report."

Peck looked down at the unbound stapled list of talking points. "Sorry, sir. I was going over options." He stood and handed over the report and watched it follow the previous document's path into shredder history.

Richard Abadon's voice had the finality of a Yale-educated business executive and Georgetown lawyer. "There are no options for you, Mr. Peck." He paused either in frustration or sympathy. It didn't matter to Peck. This case was closed.

Abadon continued in a tone that backed the finality of the statement. "My brother and I have gone through multiple options. You may have actually overheard some of it earlier. There are no acceptable options to severing our business ties with your firm's services." He bent over to slide the report into the jaws of the shredder.

"I haven't finished my report on Carlos Reyes. What we have of his backstory is revealing. Did you know he even worked downstairs for a few months? Back in 2012?"

"Absolutely, he was hired to work with Terry Imbler on our overseas security systems."

"Imbler?" Peck had heard the name in passing before.

"MacCauley's man Imbler has been following both Parrish and Reyes from time to time as we catch up with them. If he finds something…" Richard left the phrase unfinished.

Peck looked up from his fixed gaze at the glossy toe of Richard Abadon's wingtips. "But what about Tom Corliss?" Immediately, Abadon's hand rose in a palm-out stop gesture. Peck rephrased. "I have had staff losses that need to be rectified; I need follow through." Needless to say, Peck wasn't going to be filled in.

"Now if you'll excuse me, Mr. Peck. The project is closed; I think we're finished here. If you stop at Margaret's desk on your way out, she will have a termination and non-disclosure package for you to sign, and once those are signed, I believe you'll find you've been adequately compensated for your efforts."

His mouth opened in anticipation of a protest, but none would come. Without taking Abadon's extended hand, he wheeled on a heel and headed for the door. His temples were pounding; his blood pressure, never good, was rising now. He could feel the flush in his cheeks and looked for a marked bathroom door in the foyer to the two executive suites — there was none. His vision narrowed to immediate focus as peripheral vision darkened.

Christ, not another stroke! He stopped, hand over heart. His pulse and breathing were both elevated. No chest pain yet. His breathing was loud in his ears, his pulse throbbed at his neck, and his tongue seemed too large for his mouth. He saw Margaret approaching with a clipboard and an envelope. Water, he thought. "Water, Margaret. Can I get a glass of water? I need to take a pill."

Margaret stopped with the envelope offered. "Mr. Peck, are you all right?" Genuine concern pained her face. He noticed her, for perhaps with clarity attenuated by the increasing loss of peripheral vision. She was one beautiful woman for someone senior enough to wear her silver badge of maturity with such poise.

"No, I most certainly am not." He tried to regain composure. A single bead of sweat left his right temple and crawled down his cheek as he measured out a few hard breaths. "I've just had some upsetting news. I'm sure it must happen a lot up here." He looked around at the understated elegance and out through the glass end wall at the brother's shared rooftop garden now lit by subtly placed amber ground lights, then back at Margaret. He thought that a lot of bad news happens here. He extended his hand. "Please, some water."

~ ~ ~

Andrew Abadon heard the outer door close. The executive atrium door's distinctive wheeze signaled Peck's departure. He stood, stretched, and turned toward the door to the roof garden. He was glad this chapter was over: the man was dangerous. He stepped out into the sky garden, his personal label for the rooftop oasis outside his office. The amber string lights ringing the base of each potted plant or recessed seating area were originally meant to prevent guests from stumbling into them in the dark. In use, they provided just enough light for navigation on the eighth-floor balcony to allow stargazing. Planet gazing actually. He had never really learned the constellations other than the major ones—Orion, the Seven Sisters, the Big and Little Dipper, something else V-shaped that he could never remember the name of. His favorite target in the night sky was Jupiter. King of the planets, seemingly unmoving in the night sky, its twelve-year orbit kept it in view from month to month in a barely changed starfield.

A slight haze marred the clarity he wanted. Tapping his watch, he spoke into it. "Margaret, please have the parking lot lights turned off for a few minutes." A polite "Yes, sir" was the only reply needed. A few moments later the field of parking lot lights faded to black. It worked; the haze, no longer backlit from the ground, disappeared. The bright speck that never twinkled, Jupiter, suspended in its trace over five times more distant from the Sun than the Earth. He was tempted to remove the housing on his permanently mounted scope, but a shiver told him it would be getting colder soon.

He had once visited his alma mater's observatory and had enjoyed a real-time glance at the spherical moons. Returning the favor, he had given the astronomy program a million and a half dollars to purchase a new one meter reflecting mirror scope. He still remembered the names of the Jovian moons. Peering up at their mother planet, courses taken a half-century earlier tickled his memory.

He listed the four planet-sized moons: Europa, Ganymede, sulfurous Callisto, and Io. Their place in eternity would outlive the humans who named them. He knew this as sure as gravity. As sure as physics. He dropped his gaze to the flat horizon. In the distance, yellow smears, large and small clusters of humanity, shed their lost light into the night sky. Behind him, he knew, the city's lights ruined night vision for over half the horizon.

How much of that lost light could be reduced? How much of that lighting could be replaced by solar-powered LEDs? Hmm, don't we have a municipal lighting concern in Topeka? How much taxpayer budget would the city save if it no longer paid to light its city streets? Who would know that? Ah yes, the perpetually rabble-rousing Ms. Parker Parrish.

But not if she were dead. *God Damned Peck. The odious Odie Peck, thank God we're through with him.* He turned and glanced back through the glass walls of the shared foyer and into the reception area where Margaret Campbell sat, he presumed, finalizing the shredding of anything relating to Mr. Peck. A tight smile thinned his lips. He tapped the watch again. "Margaret."

"Yes, Sir?"

As she bent her head toward the speaker phone, he appraised her in profile; she was very much a beautiful woman. It had been what, 25 years since their affair? "Margaret, would you please contact Parker Parrish of Room for Tomorrow. I would like to set up a meeting to discuss funding. Yes, let her know that we are interested in participating in their fund raising, but stress that I'd like a personal interview first."

33

Andrew Abadon strode into RFT's executive suite with appraising glances left and right, as if he had expected a certain amount of faux Danish furnishings. Their absence, and the low-key replacement with comfortable living room furniture, didn't faze him in the least. He did nod in appreciation at the ceiling-to-floor water wall that covered one end of the boardroom, a flowing sheet of water cascaded and danced down a custom rock wall before disappearing into a gravel bed below floor level. Despite his reserve, his appreciation for the single slab of cypress that formed the conference table was evident as he laid his meeting folio down and let his fingers caress the table's grain. The seating arrangement had left his open chair nearest the water wall.

Parker watched as one of the world's richest and most powerful men sat down and settled into her chair, her conference room. He presented an established air of command, used to and comfortable with wealth and power. "Mr. Abadon," she said, "welcome to our offices." She and the executive board's steering committee had agreed that only she would talk beyond nicety of introductions. The steering committee, made up of her most trusted senior technical and finance advisers, was in attendance to answer questions that might arise. After introductions, they sat — some grinning in embarrassment — others subdued by intimidation, waiting as their boss and the corporate monster squared off across the table.

"Ms. Parrish, thank you for offering the invitation for an opening discussion. I've come alone, as you can see, and I had hoped for a candid tête-à-tête." He glanced left and right to dismiss the others, "No offense to your status here in Room for Tomorrow's organization. I hope you understand, and you might well be asked to return in, oh, thirty minutes or so. But I believe we need to begin this opening round, *tête-à-tête*."

All heads swiveled to Parker, who nodded, smiling. With little unnecessary shuffling, the technical team gathered their folders, rose, and filed out. Only Joanie Brayton paused at the door, glancing back with some apparent apprehension, before closing the door.

Alone now, she resumed. "Again, welcome to our place." She spread her arms inclusively.

He simply nodded an assent and opened his folio. "Ms. Parrish, there are several topics I'd like to bring up today. I suppose I might have forwarded an agenda of the things I'd like to bring to the table but there are some preliminaries." He looked up from his outline of discussion topics and, above the rim of his glasses, held her gaze. "I note that your second is not with you. Mr. Reyes?" He pointedly eyed the empty rows of chairs lining the table between them. "About eight years ago, his firm did some work for ours, as a sub to a sub, but nevertheless, it appears to might not be coincidental."

Parker's pupils dilated slightly as her danger wall went up, but she tried, really tried, to show no outward show of alarm.

He continued, "As Mr. Reyes would have had access to some of the most secure areas of our IT infrastructure, it is alarming."

He let his opening salvo linger, hoping, she supposed, for a flinch. She answered in explanation. "We, Carl and I, met over a year ago, casually, and realized we had, well, not only a personal attraction, but the beginnings of a new corporate structure. We realized we both had a deep sense of concern, no, despair, over the state of entrenchment in political and cultural attitudes concerning long-term climate trends.

"In answer to those concerns, I had the legal background to develop a corporate entity. Carl had the IT experience to build first a social media campaign, the early go-fund-me operations, and finally the data storage and fiscal solicitation and investments engine required to fully back our mission goals at Room for Tomorrow."

Abadon's eyes did not respond but his thin-lipped smile betrayed disbelief. She added, "No doubt, having exposure to the structures of your security system were part of his contract, but don't you agree that was what he was hired to reinforce? I can assure you, Mr. Abadon, that Mr. Reyes, Carl, has been my most valuable partner in helping develop our ongoing operations."

The thin smile disappeared. "Ms. Parrish, may I call you Parker?"

"Yes, Andrew, of course." She forced a smile in return, hoping it was not as overtly false as his. "And Carl is, well. He's tied up at the moment."

His eyes flinched, hardened, she thought he was seeking another response, Instead, he dropped his gaze to his folio and flipped the front outline sheet over to a second page. "There are transactions within our, let's say, secure investment portfolio that are irregular. Transfers from and within our offshore accounts have," he paused, "far too often not been completed."

"Excuse me? I don't see—"

"These transactions present on one set of tallies as transfers from one bank to another, simple transfers to be sure. But they don't add to the other account's balances. They simply disappear."

Parker stiffened in her chair; her hands dropped to her lap, one set of fingers cupping the other fist. A posture she remembered from the courts when an opposing attorney began to wander too near a hidden truth. "Andrew, it would seem that your mention of Carl's earlier work in your IT systems implies that he might have something to do with these, er, discrepancies. I don't understand. I had hoped that our meeting today would follow far different lines. We have so many projects to present, the reason for the presence of my executive steering committee. Instead, I find what amounts to this implication of theft very disappointing." She began to pull her array of talking point summaries into a pile, signaling an end to the interview.

"Ms. Parrish excuse me. I don't have conclusive prosecutorial proof that Carlos Reyes has or had any connection to these errant transactions. They do amount to many millions of dollars over a period of roughly eighteen months. That time period starts at about the time RFT, as the media call you, began its phenomenal growth. Our own IT security personnel brought to my attention this coincidence, and I need to clear this up before we can proceed to any of those further topics. It would be imprudent of me to suffer a fox to remain in our hen house."

"Your own IT security, do you have absolute confidence that they are not perpetrating these 'errant transactions?'" She smiled, a trial court-hardened cross examiner in search of a crack in a defensive wall. "I would suppose there must be a short list of personnel in your IT structure who could individually, or in concert, effect those incomplete transactions." Perfect path to a hung jury was to sow seeds of doubt.

Abadon sat on the front half of his chair, if anything, more erect, face hardened. Breaking eye contact, he looked out across the soulless landscape of downtown Anaheim's business district. Traces of the fires a few years ago were gone. Only a few blackened trees still stood among vacant and bulldozed lots. That afternoon, there was no hope of seeing the distant ocean.

Studying his composed, thoughtful stare, she said. "Andrew, my guess is that you have a much better view in Topeka. One unaffected by fifteen million automobiles." She wanted to add that his firm probably accounted for half the fuel used by those fifteen million autos, but she held herself in check.

His posture softened, only slightly. But it was a tell. He turned back to her and his agenda. "Parker, would you mind one probing question?" He leaned forward across the table, inviting the yes, but certain to proceed in any case.

Her hands, now above the table, opened in acceptance.

"In anticipating this meeting, I had expected that Carl Reyes might be here as well. I believe that I had asked for this meeting to meet with both of you. Is he still recovering from—" an uncomfortable pause belied complete understanding of Carl's injuries, "that terrible tragedy in Croatia?"

Still on guard, but sensing that the issue might be subsiding, she replied, "Carl is an amazing man, Andrew. He's working when he shouldn't be anywhere but home in bed. He is in Houston, working with a design team at Kellogg Brown and Root. One of our new projects is a levee and flood gate system for the Ganges Delta." She softened her own rigid posture. "My apologies; the meeting inquiry indicated that it would be convenient if Carl could be in attendance. I'm sorry, it wasn't convenient. We're trying to get preliminary cost estimates to Brussels, and we've been working with KBR and NGK to develop construction estimates for a World Bank hearing next week."

He turned his head slightly, chin off to one side, inviting more. "Interesting."

"NGK is preparing on-site topographic studies, some soil borings. KBR is developing the flood gate and lock system modeled on the Port of Amsterdam's floating flood gates."

"But that would cost multiple billions. What conceivable benefit—"

"Yes, it's expensive. Billions with a B. Our talks with the World Bank strategic investments team in Dhaka are centered on leveraging official support for birth control and education for all children in exchange for world bank investment in not only the hedge against sea-level rise but a wind farm either offshore or in the mangrove tidelands that will harvest nearly perpetual winds. Andrew, have you ever seen a heat map of the globe's best wind energy sites?" He didn't respond, telling a no. "Well, for the investment, the Bangladeshi get control of the tidal surges, too frequent tsunamis, and later on, protection from rising sea levels. In the deal, they also get control of population growth; education of girls is central to that effort. They also get clean energy piped in from offshore."

He held up his hand to stop her discourse. "How do you get a firmly Muslim country to agree to, to the simple act of educating girls?" He stood and began to pace. "Aren't they killing villagers for letting their girls learn to read? My God, Parker. This doesn't seem feasible from the start."

She warmed up her most winning, case-closing smile. "Bribery, Andrew. We simply buy them off with more jobs and foreign investment than they can say no too. Instead of spending millions on military intervention, we want to spend millions on economic development. Additional wind farms are planned for the drier regions inland. There are possible irrigation projects as well that will create inland lakes to store the diminishing runoff from the Ganges. Of course, we'll replace ox-driven irrigation pumps with Israeli-developed drip irrigation systems."

Abadon's eyebrows peaked in interest, so she continued. "Runoff from the Himalayan glaciers feeding the river and delta is diminishing; and those glaciers are retreating. We may see a time when the only annual flows in the Padma are from the previous winter's snowcap. We would then expect a nearly dry riverbed, and agriculture would require irrigation from those inland lakes."

"How long? Before those glaciers are expected to disappear?"

"Depends. How fast are global temperatures going to rise?"

"Hmm." He steepled his fingers under his chin, temporarily lost to his thoughts.

"Andrew, can we move formally to my agenda?"

He hesitated, turning his head to contemplate the cascading water wall. When he turned back to face her, his eyes had hardened. His briefly interested countenance was replaced by a raptor's glare. "Parker, I said earlier that I don't have conclusive prosecutorial proof of Mr. Reyes' involvement in our errant transactions. I'm putting on the table that we have sources who have a very high degree of confidence that at least eight such transactions were discovered in our offshore accounting. The amount of sophistication in creation of falsified internal records indicates a high comfort level from someone who can move freely within and between several of our mainframe installations. These are not connected installations. I'll give you this: we can begin to discuss investment options, but the issue will be pursued. If we find culpable activity on your part, or your organization's, we will prosecute."

She regretted that he could see her swallow before answering. The white noise of the water wall filled the space that should have been an answer. Her fingertips began rolling against her thigh, slowly, soundlessly. Taps that light were more like counting off points than drumming. She realized that they were still at opposite ends of the table, two frozen statues separated by a fourteen-foot-long table.

She leaned back in here chair, assuming an air of sublime confidence. "Just for the sake of following your line; You did mention offshore accounts. That brings to mind a scenario involving "offshore accounting." Suppose an inquiry was made, subject to any formal charges against any person or persons. And, in the course of legal depositions and enquiry, that the cash flows were traced. Suppose too, that the IRS were made aware of the sources of those funds, leaked on the great and powerful WIKI, for instance. If the source of these allegedly misplaced funds was determined to be unreported capital gains and profits from multiple centers of operations. What might be the implications for your many holdings, and their tax liability or liabilities? Were we talking about billions? Or was it millions? Either way, even given some clever accounting, it might be hard for the IRS to square the existence of those misplaced funds with the accounting officially reported on prior official tax statements and the related audits."

His posture began to slowly droop. If she hadn't seen it happen, it might have been hard for someone entering the room now to see anything other than a businessman sitting stiffly in his chair. But she had seen the subtle sag. Being a trial lawyer had its own on-the-job training, and her skill set in reading witnesses was invaluable in a boardroom.

Then the tell came. Abadon broke stare and glanced down at the list of occurrences when funds had been appropriated. She drove in the nail. "Andrew, I've watched an IRS seizure of accounting assets before. It's ugly. An army of toady little men in cheap dark blue suits appear unannounced with rolling carts and begin taking away laptops, tower computers. If you're lucky, and the employees have a few minutes' warning and any sense, they'll have time to power down before everything you thought you knew about yourself is being stacked in a van. Another set of guys will invade your server room with high-capacity drives and download everything they can pry into."

In hard thought now, Andrew's lower lip pushed upward. His chin dimpled in the effort. She offered a break in the tension. "Can I get you a glass of water, maybe a little scotch in it?"

"Yes, scotch with a splash, on the rocks." He momentarily looked beneath the second sheet in his folio, beyond the list of incidents of diverted funds and then flipped back to the top sheet. A list of projects he had questions about.

Parker reached across to the phone console at the table's center. "Meg? Could you please bring in a tray with water, ice, glasses and a bottle of single malt scotch?"

Meg's disembodied voice sounded surprised. "Scotch?"

"Yes, the Lagavulin, please."

"Yes, ma'am."

Leaning back, relaxed and arms uncrossed, she asked, "Lagavulin 16 OK? Or do you prefer a blended?"

"No, that's fine. Thanks." He coughed into his hand. "And the water is appreciated." The tension break had worked; Abadon relaxed into the back of the board room's low-backed plush chair. He cleared his throat. "Parker, as a veteran of many hard negotiation sessions, I've learned to appreciate a partner who comes to a meeting prepared, and knowledgeable about the changing face of business in the modern world. I appreciate your sense of understanding."

He coughed again, clearly needing a drink, but continued. He licked dry lips and gave a sidelong glance at the water wall.

The tinkling splash of water was tantalizingly close. He regained his composure and continued. "I've watched the speed with which your organization not only rose to international prominence, but at which your projects take shape and move. I understand that the Jutland wind farm is already generating power from the first two installations and that it will be fully installed in another six months. That accomplishment alone is impressive for a new organization."

"Thank you, Andrew. But you know, those plans have been permitted for over two years. They simply didn't have the match money from investors to secure the institutional funding to move beyond the demonstration phase. We put up the match and will be drawing dividends in year five. Substantial dividends. Pending our board of directors' approval, we may actually forgo those payments on condition that they serve as seed money for additional wind installations off the Scottish coast."

He smiled now. "I've played golf at St. Andrews. I'm not sure the wind ever stops there." Meg entered carrying the liquid refreshments and poured for the two of them.

~ ~ ~

To their mutual surprise, their amicable conversation over the next few hours covered projects that required either subsidy, match funding to attract and secure government investment, or pure financing on the basis of their importance. Topics covered the globe, from funding for expanding green belts of trees along the broad reaches of the savannas of sub-Saharan Africa to the need to realign Third World opinion on the importance of including girls in public education. They discussed emerging technologies, such as simply planting freeze-tolerant grasses on the thawing Canadian and Russian steppes to slow methane releases. She called in her tech team who had been waiting in reserve to discuss the numerous updated studies on glacial losses on Greenland and Antarctica.

He hadn't been aware, she found, that even though the north polar cap floated and wouldn't raise sea levels when melted, its loss of reflecting snow and ice would raise water temperatures that allowed for higher acidity and a resulting loss of coral reefs in other world oceans. Or that the resulting changes in ocean currents would likely cause at least a ten-degree drop in temperatures across northern Europe from a diverted Gulf Stream.

After an impromptu call-in dinner for all participants in the afternoon's long session, Abadon walked over to the water wall. Poking a finger in the stream, he let the stream divide noisily around the new interruption. Below the finger the stream resumed its flow as if nothing had changed.

He turned, facing the group, a more casual Andrew Abadon in shirt sleeves and a loosened tie. "You know," he said, he absently flicking water drops away, "in the long stream of things, the planet won't notice us. It will spin through the cosmos, erupt and erode, freeze and thaw, and evolve life forms we can't imagine." A number of nervous glances were exchanged at this long view. "So, I need to see the shorter picture. We may have had a short-term impact, I get that. But the planet has been emerging from an ice age for the past tens of thousands of years. What significant thing can you or I do to alter that?"

Parker was stunned. They had been talking informally for hours, exchanging views, discussing projects and their benefits. Hadn't he been in the room? "Andrew, I told some of the early members of the board, quite a long time ago, that I had dreamed a dream in which I had glimpsed a future too horrid to fully describe. It was a future aggravated by human activity and human conflict in which only a few hundred thousand individuals still existed and many of those are incapable of reproducing due to genetic damage."

As she was about to go for the big pitch, the building swayed on its rollers. Another of a series of low-order tremors passed through the Los Angeles Basin's bedrock. Glasses danced on the table. Jeremiah Eastberg from transportation innovations cried out in alarm. Those closest to the exterior window wall quickly moved to the room's inner wall or out into the hall. In less than a minute the tremor had passed, with the only change being that auto cut-offs had turned off the pumps feeding the water wall. It grew silent.

Worried that the minor quake might terminate the meeting, Parker condensed all of her arguments into one last set of statements. She excused the tech staff and sat near Andrew's chair. In the bowels of the building's tech control center, the emergency cut out for the building's environmental systems clicked back on.

300

Water began to cascade down the rock wall again, but not before Abadon reached out to wipe algae off one of the protruding black rocks. He brought his green-tinted fingertips to his nose.

"It's algae, Andrew. But you knew that. It provides oxygen to this room and helps scrub carbon dioxide from the lobby Koi pond. It's an interconnected system. This building is being converted to a low energy demand sustainable ecosystem. A small effort in our larger campaign to combat the obscenity of a hot afternoon in the LA Basin." She saw the segue opportunity and took it.

"You asked what we, you and I could do to change things. I have two things for you, Andrew. There are two things you can do that will have a significant impact that won't cost you a dime.

"First, if your lobbying efforts supporting climate-change denial groups cease and you, Andrew Abadon, officially accept the doctrine that climate change is a global emergency, public opinion will begin to change. If you are behind the campaign, it will. Public opinion has to change, to push past political posturing."

In response, his hardened public smile reappeared. But he sat again and turned his complete focus of attention on her.

She leaned forward and continued, "Andrew, if you take the formal position among your millionaire club mates that climate change is a real threat to our future, and stop funding climate-change-denying pseudo-think-tanks." Her arms rose in abject frustration. "Christ, Andrew, embrace science over politics. Politics follows money, and in large part, it's your money. Political solutions are currently hampered by your congressional caucus's fear of losing your funding." He returned a wry smile at this.

She closed that topic with one final jab. "Paris? Paris! We backed away from the Paris Accord. At least, your figurehead President did. How much did that push cost you? Welcome the United States to the back of the bus. Might as well get used to being a citizen of a second-rate power if you continue to allow such stupid responses to the real world. And *that* president! Are you kidding? Don't even get me started."

She pointed out the windows toward a darkening sky. A dim red solar disk was sinking into late afternoon smog. "That's a real world out there, Andrew, not a construct, or a market analysis, or a set of political ideals." Her eyes fixed on a miles-long row of unmoving taillights. "And those are real people out there, choking on it."

She stopped; her chest heaved harder than she would have liked. Her technical support team exchanged smiles. She had hit that one out of the park.

"Second, … no. Not yet. Andrew, I want to invite you to take a drive with me tomorrow afternoon. Carl will be back. I'd like to take you to see something that I cannot explain otherwise. You'll just have to take it on faith, that you'll come away from the experience a changed person. In short," she tilted her head at her corporate logo on the wall, "you'd leave room for tomorrow."

34

Odie Peck, still lagged from his flight in from Kansas that morning, drew his attention away from the endless waves rolling silently toward shore. The day had been just too damn long, and his night was dominated by hatred for the bastards who just wouldn't die. They had cost him the best gig he'd ever had or probably ever would have. He had them to rights, and that other bastard, Richard Abadon, hadn't even waited to hear the news. He had the required level of evidence to publicly prosecute the love birds in charge of Room for Tomorrow. He had in his pocket a thumb drive with lists of RFT's contributors. Ample evidence to bring indictments, to crash forever the reputation of Room for Tomorrow.

"Well, screw 'em all!" Peck thought. On the sun deck of his rental clifftop house a thousand feet south of the Parrish house, he fine-tuned the directional microphone pointed at her deck. He thought about the installation he'd finished in the extra hours the night had provided. Two small charges of HMX, located at critical rock faults just below the deck's steel support beams, would bring down her beloved deck and probably collapse the house. Its design, a cantilevered balancing act, lent itself to a gentle push. He had originally planned for the help of a minor tremor, magnitude 5.4 or 5.5. Happens all the time, yesterday in fact. It was on the news.

Another local earthquake like the one yesterday, and the house would go tumbling down with any residue from the explosions ground into the surf below. Just to make sure, two sets of Claymores were aimed at the sidewalls of the remaining portions built on the more or less solid cliff top.

That was before he'd gotten fired. Before he signed his house over to his mother, before he'd left the suicide note for her and a plate of cookies. He smiled over that thought. "Bitch, I'm outta here, outta your miserable life, and outta this ungrateful country, 'cept you'll think I'm dead and gone."

Then he thought, "She might not even go to the house for a while. Bitch! I don't give a possum fart whether she does or doesn't." The Caymans would be really nice and welcoming, and the final payment from the Abadons was fat enough to set up an American for years.

He picked up his binocs again and stared out to sea. Just a vacationer taking in the sights, looking at birds. Occasionally he'd pan right to check for movement in the kitchen. There was half a chance that if they were in the bedrooms, closer to the road, that part of the house might not tumble. No more almost this time. Peck had a certain conviction that somehow these two knew about Tom Corliss. What had happened when Tom was assigned to them? His last report, a brief text message, had said he was following the pair up into the hills and had just gone into park lands.

Peck had trained with Corliss as they both prepared for Navy Seal school. That neither graduated had given each a protective sense for the other, and both had used the other's skills on mercenary ventures over the years. Despite any lack of evidence, he was dead certain that Tom Corliss had not absconded with close to two hundred million dollars in a small boat into the Pacific.

His pickup mike from the Parker foyer sent a warbled note indicating a phone call to her cell phone. He bent over his gear, dialing up the volume feeding the foyer's badly echoing sound. Then he saw her. Moving through the house and into the kitchen. Yes! Alright, where's the boyfriend?

"Abadon!" he mumbled aloud. "She's talking with Abadon?" His senses were on high alert. What twist of fate could put her and her financial victim on the same phone call?

Had they pursued his lead? *Maybe put that feckless toad Harry MacCauley on my lead? Where are you Reyes? Come on out to play. It's show time!* He set the phone down on a side table beside his other equipment and plugged in its corded ear piece. Pulling a burner phone from his shirt pocket, he pre-dialed the number, leaving the last digit that would detonate the three charges. *Fuck a bunch of earthquakes! Make my own damn quake. Maybe I could take cell phone pics of her house sliding into the ocean. Send 'em to KFMB for the evening news.*

Peck stopped musing, on high alert now. New sounds were coming in from the foyer pick up. Rumbling, squeaking. What? Then he heard it, as the screech of a metal garage door stopped. The sound was unmistakable, even through the bad foyer mike: they were leaving. With the binoculars still slung from his neck, he raced into the house, grabbed his keys, and, nearly stumbling up stone stairs to street level, headed for his car.

Before starting out, he fished in his shirt pocket for the burner phone. The screen came on with six numbers dialed, waiting for the seventh. He almost convinced himself to wait, to come back and finish it as he'd intended, but he had a flight to take in the morning. Nothing was going to stop this, and why the hell not? He opened the burner phone, hit the last digit, and pressed the green call button. He thought, "For Tom." A few houses up the coast, a muffled bang sounded, followed by two lower whumps.

He waited, listening; his finger still hovered over the Mustang's start button. Should it have been louder? No, that's OK, the explosive forces would have been sent out across the surf, swallowed up, or into the house itself. Very little would have made it back. Perfect! "Too bad I couldn't get the video for the evening news." He called up his tablet's GPS tracer app and enlarged the map image. There was the Parrish car, moving toward the freeway. *I don't even have to hurry*. His black Mustang slid out onto the now-quiet residential street.

A thin cloud of sandy dust settled onto the hood and windshield when he stopped at the first cross street. A smile of satisfaction split his face in murderous anticipation of the chase. As he followed the red dot's progress, he leaned over to unclasp the under-seat holster. One-handed, he loosened the clip on the Beretta M9, letting it drop onto the passenger seat. He thumbed the cartridges, feeling the clip's spring tension: clean, full. Always check. SEAL training never goes away. With one practiced hand, he slid the fifteen-round clip back into the magazine and slammed it home.

~ ~ ~

Carl enjoyed letting Parker drive. She was probably better than he was. Once they hit the freeway system, he knew she would be able to make up any lost time in local traffic and get to the mountain house in about two hours. "Park? You're sure he's going to show up? What if all his Kumbaya talk yesterday was just posturing?"

"Well then, we'll have to start over again. But I think I've at least got him leaning our way."

"Huge turn around for someone in his position."

"It's important." Her tone was just short of strident.

"I know but—"

"Just wait. It's important that we get Andrew Abadon, or someone like him. And I'm not sure who else there would be. Most A-list philanthropes have already established themselves in given markets or have already given to us substantially. Their leanings are already well known, leftist pinko commies—tree-hugging do-gooders as far as the conservative press is concerned. But, to have one of the Abadons publicly come out with a reversal of opinion, when they have been funding crap like fracking in the Everglades, or churning mountains upside down for tar sands in Canada, or that immensely stupid pipeline from Canada to the Gulf to sell those tar sand oils to China—" Her voice had begun to rise as each new phrase was punctuated with a slap on the steering wheel.

"You want me to drive while you beat up on the car?"

"No, and you need to let me get mad. Let me burn it out of my system before we get to the farmhouse. And yeah, we need to do everything we can think of to get Andrew to see and believe what we know. Then somehow swear him to keep the covenant."

"Covenant?"

"Yes, a promise to stop being an obstructionist champion of status quo. We can probably get Chu and Grieg to help out. I called Chu yesterday after Abadon left the office to make sure that Abadon would be welcome if he got there ahead of time and fell through the front door."

"That would be something to see."

She laughed, a short bark at the back of her throat. "Hello, Andrew, welcome to a sci-fi alternative to what you have always believed."

"Park, you know that's pretty much what he's going to get anyway."

"Kind of what I'm afraid of." She sped up in a clear spot on the 101 freeway, sliding around a UPS truck and finding open road. Her dimples expressed her unspoken joy when she was moving in the neighborhood of a hundred miles an hour. They were rising toward the Cahuenga Pass through the Hollywood Hills when a habitual glance in her rear-view mirror made her take a second serious look. A black Mustang had slipped into the same gap a few hundred yards behind her Porsche.

"Crap, something's off." She let her foot up on the pedal, letting her RFT green Tesla slow to the low eighties. As she watched, the American muscle car on their tail caught up a little, but then slowed as well. Usually these interactions resulted in a fellow traveler coming up to at least wave. "There's a Mustang back there—it was behind us about that far back last time we found open road."

Carl attempted to turn around in the tight seating but the bindings on his stomach wound prevented a full turn. He pivoted to the side mirror and spotted a black dot at the leading edge of the following pack before it was obscured by a pickup. "Do you think he was just taking advantage of the holes you made?"

"No, you know, the holes don't last. It's too dynamic."

"You think he's a tail?"

"Hell, I don't know. But there's been a black mustang, pretty new, down the street for a while. The rental house."

"A Mustang? There's a million of those things out there."

"OK, let's do this. Hang on!" She kicked the accelerator, pushing the Model 3 to get ahead of a line of slower vehicles in the outside lane. At the last possible minute, she swerved across a break in the outer lane's line of cars to an offramp and flew down the narrow ramp, skidding to a smoky stop at the stop bar. Against a red light, she took advantage of a gap and took a hard left turn and pulled into the crash lane under the overpass.

"Done this before?"

She turned to him with an impish grin. "Can't outrun a cop's radio, only way out of some tickets is to hide."

She peered into her left door mirror hoping against hope that no black Mustang would appear, that she was wrong. Less than thirty seconds later, a late model black Mustang pulled up to the end of the ramp. To make sure of the image in the mirror, she turned in her seat and made the positive ID.

"There, that guy who just pulled up, looking at us. Is that Peck?"

He turned in his seat, pulling against bandages, and craned to see out the rear side window. "Son of a bitch! Yes, that's Peck." He could tell even behind mirrored Ray-Bans. The military hair and fleshy jowls. "That's our guy."

She felt a tingle roll under her scalp and scratched at her left ear. Even with a now green light opening its lane to traffic, it stayed in place. The driver was staring back at her, smiling. Another car behind the Mustang honked and honked again with more insistence. She heard the tuned exhausts grumble loudly, and with a squeak of its tires, the Ford pulled out, turned right, fishtailing, and sped off away from them.

"Way to go babe. You called it; you were right. Now what?"

"We go to the farmhouse. We've made him. I don't know who he's working for, but Mr. Peck has our number. We're going to have the car scanned again and every time it comes to a stop if necessary. I don't have time for this crap. And I don't want to be out here in the world next week dueling with Mustangs in a Tesla."

Without saying another word, she took advantage of a break in passing traffic. With tires screeching she pulled out, tweaked the wheel to get the car to turn a 180 and, with only one car to wait for, took a fast left onto the freeway on ramp. She emerged onto the freeway moving too fast and had to slow down, both the car and her nerves. This would not be a good time to get pulled and towed for excessive speed. Carl reached over and placed his hand softly on her thigh.

"We good?"

She was biting her lower lip to the point of pain. She quit but continued to stare into an unfocused vanishing point. "We're good." She sighed, wondering if the bastard Peck was off their backs for a while. She turned to meet Carl's concerned eyes. "We're good. Let's go meet Mr. Abadon."

35

Peck came to the end of the ramp pissed. The two lovebirds were mocking him. He could easily see the eco-green Tesla parked on the shoulder under the overpass, Parker driving. He sensed a smirk in her smile as her sunglasses turned first to the side-mount rear view mirror and then swiveled around to see him. Was that a finger? *She shot me a finger! Bitch!*

He noticed the insistent honking of an impatient driver on probably the third honk. The chase had consumed him. With a prominent screech of tires, he made a hard right and accelerated to the next intersection. A local fast food driveway offered rear parking cover and easy access back into traffic. He checked his hand-held tracking device, pulled at the screen with thumb and forefinger, and zoomed into the immediate neighborhood. The blinking red dot indicated the tracker on Parker's car had just re-entered the northbound ramp back onto the freeway.

A glance at the food offerings on the drive-through gave him pause, but his decision was easy. He was far too hyped up to eat. He gunned the Mustang, made a too-noisy exit from the fast food restaurant's parking, and was soon on the freeway roughly a mile and a half behind his target. He noticed his hands were damp on the wheel. Breath close to wheezing. Slow down, no hurry, he thought. I know where you are, and I think I know where you're going. Traffic was moving easily for a change. He closed the distance to a half mile, keeping a few vehicles between them at all times. The roadway made a long sweeping turn to the left and began to slow; all three lanes showed flashes of brake lights as the Golden State's traffic joined the 101.

He checked the scanner in frustration, the red dot was moving. "Jesus!" he yelled. He bashed the steering wheel repeatedly, "Aaahhh!" He wanted these two meddling shits gone. He wanted to know they were damaged or dead before he left the country. But the three lines of traffic ahead were congealing into a slow-moving ooze. If there was an accident ahead, he knew he'd soon be at a standstill. Turn signals, smiles and the illegal use of a blue flasher pulled onto his dash secured access to the outside emergency lane. It was blocked by an aging pickup with steam rising from its hood. He was blocked in a four-lane, four-mile-per-hour, smoking metal ooze that was gelling to a stop. Craning his neck above the line of nearly stalled lanes of vehicles, he saw the small familiar outline of the Tesla squirt out onto the emergency lane and accelerate away from him. Despite his own illegal maneuvers, he cursed them for their disregard for the law.

He reached for the dash and grabbed his emergency blue strobe unit. With pistol in hand, he leapt from the car, ran ahead, and stopped at a silver Camaro in the outside lane two cars ahead of the stalled pickup. One good knuckle rap on the window got the startled driver to look at his shiny gold PI badge. He closed it before she could read any detail. The window came down. "Police pursuit, ma'am. I need to requisition your vehicle." In a sputtering state of confusion, the twenty-something driver was convinced by Peck's firm grip on her shoulder to exit the driver's seat. He tossed her the keys to his Mustang, nodding back down the traffic line, "This car's just a few cars back, black Mustang," and told her to get in touch with the LAPD to exchange cars later. He was running on advanced adrenalin now, and no doubt his adrenaline-induced mania helped convince the Camaro's driver of his real urgency.

Peck pulled the Camaro onto the emergency lane and peeled off in pursuit. He glanced at, but barely noted, the accident to his left. A panel delivery truck had rear-ended a nineties-era Town Car. The two vehicles blocked the middle and outside lanes, and a slow-moving line of cars filed by in the remaining inside lane. A stoop-shouldered, blue-haired woman stood beside her car berating the truck's driver. Ahead lay a long stretch of nearly open road and, far ahead, Parrish's goddamned Tesla. He floored the Camaro. Trailing a cloud of blue-tinged tire smoke, he rapidly pushed his speedometer's needle past the hundred-mile-per-hour tick mark. With his blue strobe pulsing, the few cars that had made it clear of the accident scene scrambled to give him clearance. The California Highway Patrol employed a number of confiscated muscle cars as pursuit vehicles, and his Camaro fit the bill nicely as a CHP chase car.

Ahead, Interstate 5 crossed under the Reagan Freeway. Shit! Where would they go? He remembered that Tom Corliss was on a trail up into the hills. He pressed the pedal down, passing and weaving with reckless intent to catch them or die trying. Again, the freeway slowed, this time as the 405 merged from the south. Nothing for it. He took the inside bail-out lane and charged ahead. He finally cleared a bottleneck and the Camaro surged ahead. In two minutes, they were in sight.

Peck decided to continue the ruse and pulled up to within twenty feet of Parker's Tesla and flashed his high beams. Both cars were moving at well over eighty miles per hour. He flashed high beams again three times as he'd seen the CHP signal traffic offenders. At first the two visible heads bobbed to look at rear view mirrors and the car began to slow. Were they uncertain? Had he fooled them? He put his left fist over his mouth, both to obscure any facial features not covered by the sunglasses and in imitation of a hand on a radio's mike.

The Tesla slowed and veered to the shoulder, slowing but still moving. Peck pulled to the left and alongside the smaller electric car. Sending a one fingered salute in greeting. With some satisfaction, he saw Parker's look of recognition and then horror as she saw his car veer into hers. He slammed into the side of the lighter car. The crunch of metal on metal was gratifying. He veered left and then right again, but on the next attempt to body slam the smaller car, he found air and swerved on the shoulder's loose gravel. Her car had abruptly shifted backward, out of his view. The Tesla was now in his rear-view mirror and swerving into a side spin. He watched as it crossed back into the traveling lanes, reversed spin, overcorrected, and headed for a grassed embankment. His last glimpse of it was a dark shape sliding sideways toward the pilings of an exit sign. He took that next exit.

Peck felt the pulse in his temples, noted his sweating palms, and knew he needed to dial down a notch or more. The near faint in the Abadon executive suite had only been days earlier. He took a right turn on the cross street, a four-lane feeder, and pulled into a strip center parking area. "Shit, shit. Shit!" He took a deep breath and another, attempting to slow his breathing and, hopefully, to slow his pounding heart. His mind was still racing. Did he really just do that? Did he actually just steal a car and attempt vehicular homicide? His pulse thumped in his ears, his forehead glistened, and a damp sweaty stink began to fill the car. He reached for the glove box and remembered. *SHIT, Camaro! My heart pills are in the Mustang.*

Slow down. Slow. Think. Think! *Far too many people were witness to the carjacking, not to mention the car's owner would certainly be able to ID me.* At least a half dozen other drivers would be able to give a description of the chain of events. He ran through the options: *I still have airplane tickets and a passport in my coat pocket. I could head south to LAX and be gone, away from LA forever.* Caribbean beauties waited for him, holding umbrellaed drinks. Most of them he knew just wanted to settle into a bungalow with a rich gringo. *I want to be that gringo, and I'm rich enough.*

He felt his chest, then his jugular. His heart had slowed, the adrenaline's burst had left him, if anything, a little tired. Scanning his surroundings, he absorbed the bucolic suburban scene: citizens in shorts and T-shirts, shoppers of all sorts, an ancient pushing a walker up a ramp. All going about their business in a mini-mall carved into the hillside. He closed his eyes and instead envisioned a palm tree-lined pool. God, he wanted that life. "It's only a day away. Just one more detail!" He reached into his coat pocket for the tracking device. The red dot was stationary, a quarter mile southeast of him and a little bit to the right of the roadway. He stared at the device, thinking.

Tapping the menu to pull up its history, a scramble of red dots filled lower Los Angeles County, Orange County, and San Diego County all the way down to their now destroyed cliff house in La Jolla. He laughed silently. *Now it's a beach house.* The laughter erupted into a full bellied guffaw; his tensions seemed to fly away. He'd fucked them up and good. House gone, fancy electric car gone, possibly they were gone. He wiped a tear away with the back of his hand. There was one line of dots — actually, several overlapping lines of dots — that went farther north into the hills.

"Where the hell is that? Dammit, I should have checked that out!" He looked up from the little device. "What are you two bastards hiding up in the hills?" He silently mouthed a string of curses. "Is that where you keep your secrets hidden?"

The pulsing blue flash of the portable police strobe pulled him out of his distracted reverie. He flicked the off-on switch and glared at two fast food customers who were crossing in front of him, staring. He looked back at the little display, curious now. Did they have a canyon house too? Should he have one more go at discovering a provable link between Room for Tomorrow and the theft of millions of the Abadons' personal fortune? *Could this be their secret stash? Over two-hundred fucking dollars!* He poked at the farthest extent of the red trail, zoomed in again, and saw that it was in the hills, valley side. He set the hand unit down and put the car in gear.

~ ~ ~

An hour and a half later, Peck crouched in the fresh green underbrush that had taken over the fire-blackened hillsides surrounding an oddly out of place two-story house. A damp winter and the naturally regenerating vegetation found in wildfire country had created a waist-high riot of diverse greens. In brilliant mid-day sunlight, he hoped his ill-chosen street clothes were hidden enough by the low brush.

The hijacked Camaro was hidden behind a boulder pile two switchbacks down the road. He wondered briefly if the islands extradited for grand theft auto, then decided he didn't care. The false identity kit would settle all accounts in the US. He didn't even mourn the loss of his beloved Mustang. He thought more of the battered Tesla in his rearview mirror and the Beretta in his palm. Fifteen rounds ought to be more than plenty. For what? Could there be someone in there already? He started to rise to mid-height when chopper noise suddenly erupted from the other side of the hillside. He flattened and waited for it to pass over. No need for anyone to witness a person hiding at the scene of the crime.

What the hell? The chopper banked around and landed just clear of the narrow lane gravel park road. Slowly rising above the greenery, he stared in jaw-drop wonder as a sleek gold-and-black Bell corporate chopper powered down. The pilot, dressed in a traditional dark blue and white, was visible through the windshield flipping switches in post-flight shut down. The turbine whine died completely but nothing happened. Dust kicked up by the slowing rotors settled. The pilot, now out of his seat, had opened the passenger slider and lowered steps in preparation for his passengers.

Peck trained his field glasses on the dark interior. He was crouched in an uncomfortable forward-leaning position, restricted

313

by sharp rocks and bracken into a precarious perch on a small flat rock. He rotated the front polarizing lens until he could see through the glare on the chopper's windshield. The appearance of Andrew Abadon stepping out of the chopper startled him, so much that he slipped forward, almost falling into the brittle new growth. How, what? Why? He was pretty sure the Room for Tomorrow crew and the Abadons hated each other, so why?

He settled back on his haunches, pleased with himself. He'd finally followed up on the northern extent of Parker's travels and stumbled onto what? *Conspiracy?* He sat back, mind whirling, seeking solutions to an entirely new puzzle. *Can I still get away with decapitating RFT? Should I bag it and leave? Why is he here? And now? This is a seriously out of the way meeting place to discuss…what?*

He found a spot a few yards closer where he could rest on elbows and use the binocs on the house and chopper. No more activity from the Bell or house. No, wait…the pilot returning from the other side of the house, looked once back over his shoulder, and proceeded unhurried to sit on the floor of the open door. Focusing in, he watched for any clue. When the pilot reached for a cigarette and lit it, he moved his binoculars to the house. Odd? Downstairs, he could see into the lower floor's barren rooms. But not the front upstairs windows.

Movement! Downstairs closest window. He could clearly make out a man descending stairs and encouraging Abadon to follow him. His eyes strained for information but got nothing as their feet ascended away from him and to the second story. From the angle of the staircase, he expected movement upstairs and panned the several south-facing windows. Most were darkened, but focusing beyond the glass, his lenses allowed a limited view of the interior of rooms that appeared to be as empty as those downstairs.

The front set of windows was also darkened, no visibility beyond the glass. Peck's right finger experimented with the binocular's outer ring that adjusted polarization, and the view only got worse. Something in the view was shimmering and he couldn't clear it. Now, on maximum polarization, he could clearly make out a pulsating shimmer of light from the side and front windows of the upstairs front room.

He blew a puckered silent whistle. "Weird…what the hell is that about?"

Scanning back to the chopper, the pilot stood, mashed his smoke out in the gravel and climbed in. Odie took the chance. He rose to a waist-high crouch and moved quickly into slightly higher brush downslope and closer to the house. One quick glance for clearance showed the pilot pre-occupied with a clipboard. He heard jazz sax coming from the onboard radio. Peck rose and sprinted for the side of the farmhouse. He moved toward the back of the house and, peering around the back corner, stopped in shock. Tom Corliss's car! Tom's car, an abandoned wreck on blocks, advertised the man's disappearance. Tom Corliss had come here on his own and vanished. Where?

36

Chu nipped at the nail on his index finger. He only chewed his nails when he felt useless in the face of things to do. CalStation would be getting crowded again and they were to expect at least one more visitor from the past. He thought about the only times there had been five people in the small space. Over a year ago when Parker and Carl had first arrived. Noel had been alive then. He turned away from Grieg and away from the spot on the floor where Noel had died. The other time had been months ago when Tom Corliss had made a one-way trip through the portal from 21st to 23rd, never to be seen again. Both times had precipitated changes, and someone had died. The most recent text message from Parker said to clean up, look presentable. Andrew Abadon was due in a few minutes.

Abadon had been vilified for most of the months they had shared conversations with the two friends from the 21st. Now helicopter sounds could be heard approaching. Their few eating utensils were cleaned and put away. It was as clean as it had been when their self-appointed housekeeper, Noel, had kept them organized. It had taken several hours and Carl's help hauling away bags of refuse to bring the portal to its current level of presentability.

"Grieg," he said softly, "how's the presentation going?"

Grieg shrugged. A noncommittal gesture. Grieg didn't get excited and he didn't rush. The go-with-the-flow team member, as Parker had pegged him. Chu tried again. "They are about to walk in; is it ready?"

"Cha!" Grieg responded a little too loudly for the small space. "I've put in enough from the Pale Horse Archive for him to get the point. I added the Last Day." He rolled away from the desk top a few inches and waved the screen closed. "So, are they supposed to come with Mr. Rich? Didn't I just hear a chopper landing?"

Chu frowned. "I didn't think so. I thought—" As soon as he looked for it, his cell phone chirped. He glanced at the message board from Wellington. It was blank. He picked up his cell phone. The phones supplied by RFT had been handy for communicating in the 21st, but his latest message from Carl was ominous; [delayed in traffic, BE CAREFUL]. He picked it up, reading again. "What the...?" He read it aloud to Grieg. "Fat Cat comin' no time down, an we s'pose fake it? Fuckit!" He silently mouthed, Shyte! "He put be careful in all caps!"

Grieg turned from his workstation. A flat expression showing nothing, raised eyebrows passing the duties away. "Better go be a welcome mat, manno. Da Fat Cat ain' goin ta comprend wid dis on he lone."

"No, and we better stick with da 21st, no man?"

"Yes, sir. Like I read you, man, like." Grieg exaggerated Valley speak.

"No, do it straight. You know?"

"Gotya, man, kid you. Sabe?"

Grieg took a few more keystrokes, saving the file to presentation mode, and stood from his usual slouch over the terminal's manual input pad and rolled his neck, sinews popping. He said, "Good enough, convincing I think, but not too polished."

Even through the field buffer, Chu could sense heavy footsteps in the farmhouse. The proximity buzzer for the doorway to the 21st sounded. Chu shook his head, a little jealous of Grieg's disengagement, and headed out to intercept the Fat Cat, Mr. Andrew Abadon.

Chu heard the unmistakable squeak of the kitchen hall door and hard leather heels on the wooden floorboards in the kitchen. He called out, "Welcome. Coming down to me you."

The footsteps in the kitchen halted. Chu reached the bottom of the stairs near the front door and saw a tall, well-dressed man in business casual. Open collar, blue blazer, polished shoes. Chu stood at the bottom of the stairs facing down the central hallway. "Welcome, Mr. Abadon. We're just upstairs here." He gestured with his right hand at the newel post.

"Is Ms. Parrish here?" Abadon had stopped at the doorway to the central hall, appearing hesitant. "There weren't any cars. Well, there was a junker, but…"

"No, I just received a text that she was delayed in traffic." Chu put on an unfelt smile, "I don't think you had that problem."

Abadon began slowly at first, and then, with a show of confidence, he strode out of the kitchen toward the front of the house and extended a hand. Chu took it, noting the firm grip. "A pleasure, I'm Solomon Chu. Come," and he turned to the stairs.

The upstairs hall wasn't well lit. Its sole illumination was by the light of the side windows through open bedroom doors. "I came out to help you with the doorway." He nodded over his shoulder to the open door at the front of the upstairs hall. "What has Ms. Parrish told you about this farmhouse?"

"Only that it was more than it seemed." Abadon looked around the unlit hallway, a dismissive notch in his upper lip suggesting but not quite expressing a sneer. "It would not have to be much at all to be more than it seems.

"Mr. Abadon—"

"Andrew. Please call me Andrew."

"OK, Andrew. I'll go through first demonstrating how to step through. The last thing I want is for you to hurt yourself."

Abadon looked beyond Chu toward an open door. From his position, he was able to see past Chu and through the front room's windows to see the last slow revolutions of the rental chopper's blades. The pilot sat on the sill of the open side door, smoking. There was little for him to see that warranted any sense of danger. "I think I can do just fine. Are there loose floorboards to worry with?"

The bald Oriental said, "Nothing like that, sir. But it is important to watch your step. Just a mild warning. Stepping through that doorway can be a little disconcerting. You will see a small flash of light as I step through. Nothing to worry about." He could sense both wary respect and condescension in the young man's tone but decided to see, at the least, what was in the little room. He turned toward the door and watched in mild amazement as his hand activated a shimmering gold curtain.

Abadon faltered back a step, raising his hand as if to ward off danger or attack. Chu tried his most soothing tone. "Mr. Ab—,

Andrew, I assure you, nothing will harm you. You can't understand how much I value the trust you have placed in Ms. Parrish to come here and in me to follow me through this forgotten little farmhouse. In a moment you are going to feel as if you've stepped into a science fiction movie. But I assure you, it is necessary for you to believe the reasons…" He felt he shouldn't get too elaborate so soon. "Reasons that were good enough for Parker Parrish. You really do need to follow me through this doorway."

"But what did I just…?" Abadon's mouth sagged a little after the last syllable, not certain he had registered what he had seen. He searched for some belief system that allowed him to square with Chu entirely disappearing as he stepped through the front bedroom door, in a flash of shimmering gold that dissipated immediately as Chu cleared the threshold. Beyond that door, the chopper still sat in the sunshine. The pilot, he could see, was flipping through a magazine.

Chu had stepped through the doorway and had come to no apparent harm. He followed, stepping into a small room almost entirely filled by a small stainless-steel table and what looked like high-end cabinetry. Chu had continued to the far side of a small alcove and stood, ready to catch if Abadon had stumbled as some of the few visitors had. Cabinets seemed to cover each wall. A smaller man, possibly Slavic from his square face and sandy hair, stood smiling. Abadon moved closer, curious. His face transformed through wonder and astonishment to registering a reality. Chu had been right. The simple farmhouse he had entered *was* far more than it seemed. The portal was designed for compact efficiency. Its clean lines conveyed functional design.

Abadon realized Chu was talking, "…and this is Grieg, formally, Griegor Grigoryev." The smaller man nodded. His smile was unreserved and welcoming. He reached to shake the multi-billionaire's hand. Abadon took the extended hand and shook it, receiving a firm confident grip. Abadon looked toward doorways to the left and right that led off to other chambers. He compared his memory of the small room he had seen at the end of the hallway and realized it could not possibly have held this much space. A touch of bile reached the back of his mouth. With a grimace he swallowed.

The small man named Griegor spoke up. "I've seen that look. Let me get you a glass of water." Grieg turned into one of the two side rooms that connected through oval framed doorways left and right.

Abadon swiveled his head, taking in details, some familiar, many not. Function was function. Fixtures were unfamiliar, as if he were on a different continent. Outlets looked wrong, but recognizable in function as power plugs, just foreign. Countertops were opaque black but for a few rectangular areas that glowed a soft, clean white. In the corners of each of the white panels, indicator lights blinked slowly, revealing, he supposed, some status condition. The ceiling panels glowed a uniform yellow-white, as if the illumination was intended to imitate sunlight. His searching eyes drifted to a small pin-up fixed above a workstation that portrayed a voluptuous woman in full coverage gray body paint with streaked orange hair and nipples highlighted in the same orange. Her bright red lipstick a perfect circle. The image's unreadable caption in Asian script, he thought Chinese. He asked, barely audibly, "What is this place?" Then, "Where is this place?"

Grieg had followed Abadon's gaze and got up to remove the pin-up photo.

"Actually, the question should be *when* is this place," the taller man named Chu answered. "Mr. Abadon. This will come as a great shock to you, and I apologize for that. Parker was supposed to be here to help with the explanations. You are absolutely safe here." The Asian man shrugged. "We've been living here for over almost two years." Chu extended his arms expansively, inclusive of their surroundings. "You are standing in a room that occupies the same space as the upper front bedroom of a small farmhouse in rural California. But it is not of the same time. You have just stepped out of the 21st century and approximately 200 years forward into the 23rd century."

Abadon turned, his motion abrupt, surprising himself in his own gut reaction, his need to leave. He oriented to put the doorway behind him. The executive chopper should be through a window directly in front of him. He turned again, looking back. The doorway he had just stepped through no longer looked out onto the central hallway of an abandoned farmhouse. The rectangle shimmered. The view through the wavering energy field reminded him of a blend of aurora borealis and moonlit water. The view was not through the doorway anymore. He absorbed the last words

Chu had just spoken and slowly turned back to face the two men. "23rd century?"

"Yes, sir, Mr. Abadon." Grieg had spoken through a wry smile. "We're not from here, or rather, not from now."

Grieg handed Abadon a small stainless-steel tumbler of water, took a small sidestep, and opened the doorway to the portal's air lock, decon shower, and equipment and changing room. Then he backed into it to make room in the small antechamber that served as the four-seat cafeteria and common area. Chu stepped to the other corner and slid open a panel that provided entry to four small over-and-under bunks.

Chu prompted. "Mr. Abadon, we would have waited for Parker and Carl — er, Ms. Parrish and Mr. Reyes to arrive." He paused to look at his cell phone. "But they were delayed, as I'd said. Grieg has prepared a slide show or video, a combination actually. A video, which will make a lot of things make sense." He motioned for Abadon to sit at a workstation that appeared to have no controls.

Grieg pulled a small remote from his pocket and keyed the workstation to begin. Abadon flinched back slightly as a display, previously invisible, began to glow. As the holo-screen solidified, images began to flow depicting fantastic architecture. He identified iconic structures in the cityscapes that anyone would recognize; the Freedom Tower in New York, or the Shard in London, Paris's Tour Eiffel, Beijing's CCTV tower, or Tokyo's Skytree. Common to all of the scenes were additional buildings that matched or exceeded them for architectural innovation and height. Towers connected to towers at intervals by seemingly spun-glass bridgework. Impossible extrusions held landing pads or gardens in the air.

The view changed to earth orbit. As he watched, complex space stations appeared as a visitor would see them on approach. The Hilton Drum Hotel depicted in eclipse with hundreds of glowing rooms facing the viewer, a brilliant lens-refracted flare of the sun peaking from an edge. The Marriott Wheel and the SISS. The Second International Space Station's rabbit run, reminiscent of Kubrick's *2001* space station, was displayed in flyby. A view of Lunar Habitat II was shot from a camera mounted on the space elevator that permitted transfer of bulk building materials into very low-G for shipment to an orbiting smelter. The images were fantastic, futuristic, and Grieg's minimal addition of text and dates in explanation provided the context for images that would be unknowable in the earliest years of the 21st century.

The graphic parade ended with a simple set of line graphs showing population in billions with plateaus, dips, and surges. Grieg had let the view of the chart back away, allowing for additional population to grow from the 1900 to 2000 scale to include new centuries. Some of the dips he'd labeled: 1918 WWI and Flu epidemic, 1945 WWII, a minor dip in the early 2020's marked the COVID epidemic. The chart's labeling continued with the 2028 1st Turk-Arab war, 2048 N-Virus, 2063 2nd Turk-Arab war. Two of the rising population sections were labeled as well—algal protein farming and herd limits on beef grazing. At the Amblyomma, the 2087 epidemic, the curve dipped by about fifteen percent. And never recovered, dropping in steps and stages to the precipice. At 2153, Last Day, the red line plunged so close to the bottom of the chart that it appeared to go to zero. Grieg's video zoomed into the bottom and a number, 489,000,000. The camera view then again backed away, but the red population line only hovered around the half-billion line. The line stopped a few tick marks into the 23rd century.

Abadon straightened in his seat and turned to Grieg. He shrugged, dismissive of the dire prediction of the red-lined population curve. "So, I've been brought to this very clever set and shown a what-if scenario?" He was about to continue his protest, but Chu stood, legs apart with both hands in a stop gesture.

"Andrew, please, there's more."

37

Abadon's mouth shut. Chu looked down to Grieg and nodded. Grieg waved two fingers at the holo-screen again. On the black screen, a simple title scrolled: "The Last Day."

What followed was a set of lime-lapse images; a view of Earth from an array of geosynchronous weather satellites. The images showed the planet centered approximately over the Korean peninsula. Most of the planet was in night, and a golden spider web of light showed the spread of humanity across eastern Asia, and in dawn's light, a bright blue arc lit a portion of the Pacific Ocean. The camera's built-in optics shunted the sun's glare, but as the time lapse sequence began to move, a brilliance flashed near the dawn line, followed by three more in the same general area. The nuclear exchange had begun in the far eastern hemisphere. The locations of flashes were indistinct at the edge of the center of the view. Flashes spread throughout the morning of the last day. Skipping ahead of the advancing dawn line.

By the time the sun's early light cast shadows in London. That city and most of Europe's major cities flashed white as the New Soviet State took preventive measures against ETO forces. The European Treaty Organization had already launched against the NSS, and growing circles of bright orange pressure waves bloomed in the old Russian heartland. In the still-dark new world, thin contrail tracks of ICBMs could be seen erupting from numerous spots well-inland and from polar locations that could only have been submarine launches.

Abadon raised the steel tumbler to his lips to wet a suddenly dry throat. The horrid little circles began to appear across the eastern seaboard. The white, then crimson, then yellow pressure waves spread out, merged, and marched west to other major population centers and military installations. He couldn't be sure if one of the bright city spots near the western edge of the globe was

Topeka, but it didn't matter. Most of them were obliterated by their own little blooms.

Abadon nearly dropped the tumbler. A cold splash of water on his ankle brought him out of rapt of attention, but he continued to watch as the uncaring sun finished its march across the dying planet on what would be called by its survivors The Last Day. The time-lapse sequence finally stopped as the sun's arc slipped around the western horizon.

The three inhabitants of the manned weather laboratory had waited until they could be sure of finding a receptive station on earth to receive the full fourteen-hour record, proof if any world court survived, of who shot at who and when. It hadn't made any difference.

Softly, Grieg provided an epilogue. "Sir, this is the only record known to exist of the sequence of the last day. It was bounced to the UN's station and the various consular offices attached to the Hilton and Marriott."

Abadon's head jerked in surprise, "A Hilton? Which Hilton?"

"The orbital Hilton, it was called Conrad Station."

Abadon closed his eyes in thought, reviewing what he had just seen. The scenes looked real enough, but... "OK, for argument, who took those pictures? The ones from space?"

Grieg answered. "Sir, there were crew on a manned Earth Sciences laboratory. I don't think you really understood what I just explained. They filmed the whole thing. This time lapse was edited down for viewing, fifty or so years ago, because the real one takes fourteen hours to watch, and most people can't do it."

Abadon tried to check his skeptics hat at the door. Could he get around the visuals he'd just seen? How to understand the magical transformation of the front bedroom to studio set? He was on the outskirts of Hollyweird. Almost anything was possible in this part of the world. "Wait a minute, you said fifty years ago this film was made by people in a manned weather satellite?" He coughed into his hand. "Fifty years ago we were just learning to make good use of CinemaScope and Technicolor. There aren't any manned weather satellites!" His voice had begun to rise, and he checked himself.

Grieg took advantage of the pause. "I think you're thinking fifty years before *your* time." He pointed to the dual time clocks on the wall. "Fifty years ago, in my time, the civilized remains of humanity were trying to figure out what just happened—who did what when. Not that it matters much."

Abadon looked at the two sets of numbers, the two second indicators rolling in unison, but the two-year marks inexorably separated by 194 years. Easy trick, he thought, any stage manager could pull that, and whatever they did with the door, the alternate view?

"You said the civilized remains."

Grieg shrugged. "There are places that have become very dangerous to be without a significant armed contingent. Places with not enough radiation to kill you but, think of any of the post-apocalyptic scenarios you may have seen."

Abadon nodded, thinking of a few—some had zombies or AI masters—sci-fi. His thinking was interrupted by a thin beeping. Chu turned and tapped at a red blinking button in a field of buttons on a virtual desk display that had come to life on one of the white areas on the countertop.

"Parker?" Grieg asked Chu.

"Mebbe, too she said be careful. Member? She said in caps!" Chu's eyebrows raised in uncharacteristic caution.

"Cha, Naprob, manno." Grieg retreated to one of the side compartments. Abadon heard mechanical clicks but couldn't determine what was going on. He'd become distracted by the shorthand English he'd just heard. Decipherable but odd. He watched as Chu's fingers danced in front of the virtual desktop. *Nice effect! They are really trying to put it to me.*

Chu had pulled up a small screen display, no larger than a pack of cigarettes, that revealed Odie Peck at the other end of the hall, gun extended in a two-handed grip. Alarmed, Abadon said, "That man is dangerous, a possible psychotic," and began to move to cover in the side room that housed the four sleeping pods and wardrobe storage.

From his vantage, the leader of global industry watched as Peck made slow, careful moves down the hallway. He wondered if Peck had a view of the valley or this stage set. Peck's moves, quick glances around doorways, fast steps in, gun first, bespoke his SEAL training. Abadon's neck hairs prickled, as Peck methodically checked the two side bedroom doors and proceeded to the entrance of the set. Chu stepped to one side of the entry door and crouched, muscles tense. Grieg pulled Abadon gently around a corner, out of the line of sight from the hall door.

Sotto voce, Abadon said to Chu, "Be careful, he's a SEAL," then added, "Navy commando." In the few moments Peck took to clear the side rooms and approach the front bedroom door, or stage set, or whatever he was in, Abadon considered why he had just sided with these two characters and not his former investigator. Peck was dangerous? Peck was possibly angry at him for being summarily dismissed. Why was Peck even here?

Peck's gun and two hands entered the space inside the field outlined in a shimmer of energy and withdrew. A guttural exclamation spiced with unintelligible curses was heard, clearly now. "Fuck it."

Peck charged the doorway and received a hard kick from Chu, who rising from his crouch, unwound and delivered a punishing roundhouse kick that simultaneously loosening Peck's right hand from the gun's grip and nearly crushed his neck. The kick, combined with the temporal disorientation of passing through the gate, laid Peck out on his back. Chu leapt, reaching for the gun still in Peck's left hand, and managed to hold it with his two hands pointed up. Grieg returned to the foyer with a sidearm of some sort. Abadon couldn't identify it. Its high, keening whine reminded him of a high-speed turbine, maybe, or charging crash cart paddles.

Whatever it was, Grieg pointed the thing at the struggle on the floor. Grieg shouted to Chu, "Chu, roll! Got im!" Chu rolled to the side and Grieg fired the thing. Flames and spark erupted from Peck's left shoulder and his gun dropped. Chu retrieved the handgun and scuttled away, getting enough space between himself and Peck to stand and maintain cover on the intruder.

"Good shot, Grieg!" Chu stood, making a quick glance to Grieg, acknowledging the hit, and then over to Abadon. His hard stare asked the unmistakable question. "Is he with you?"

"No!" Abadon leaned into the foyer space. "I fired him."

"You sure?"

Abadon felt heated animosity roiling off Chu. Defensively, he said, "He was paid off. He used to work for me; was hired to see if your friends at Room for Tomorrow were stealing from me." He gave a helpless, hands open non mea culpa. "I have no idea why he's here. I thought we'd paid him enough to go away. He's caused enough trouble."

Peck groaned and waved his arms, confusion apparent in his flailing search for understanding. His right arm reached to his left, feeling for the pain. The stink of burnt flesh and melted rayon filled the small room. Tendrils of smoke drifted toward the ceiling air recycler's intake.

Grieg knelt quickly and looped white cordage around Peck's ankles and a table leg. The cord shrank as Grieg released it, drawing Peck's legs together and attaching him firmly to the small table's leg support. Peck moaned and began to struggle against the tightening leg restraint. The table leg, bolted to floor and table, didn't budge. Peck curled into a fetal curl and seemed to pass out. Chu reached down and slapped Peck hard across the face. "Wake up!" *Slap.* "Hey, arsehole!" *Slap.*

"Owww!" Peck's face turned up. He squinted against the brightness of the ceiling panels, then searched for his tormentor and found the contrast-darkened form of Chu leaning over, his arm raised in threatening sincerity.

"You're an uninvited guest, wanker." *Slap.* "Private invitation only." *Punch.* Peck blacked out.

~ ~ ~

Peck was neatly tied, bundled, and only occasionally conscious when Parker and Carl arrived. Abadon was saying his goodbyes and would have left but for the urgent text message, then phone call from the highway that they were at the foot of the mountains and only ten or so minutes out. Abadon had stayed, sharing with Chu and Grieg his brother's prior business arrangements with Odie Peck. The proximity alert in the outer hallway sounded as Abadon stood to leave.

"One sec, Mr. Abadon," Grieg said.

"I've been here far too long already. Though, I can't say that I've enjoyed all of it." He waved a hand at the bundled lump of Peck in the corner by the farmhouse door.

"They'll be here in a second; that sound means they're downstairs." Grieg gently pulled Abadon away from the entrance to the farmhouse.

This was proved a second later when, with light flashes of energy, Parker and then Carl slipped through the shimmering boundary at the doorway.

After they passed through, Abadon continued to look at the sparkling silver curtain that remained when the physical door to the 21st was open. "I think I can explain to myself and anyone interested how you've done almost everything I've seen today." He paused and wondered if he could actually explain the energy bolt that disabled Peck. But he continued, "How do you do that with the doorway? It's really very convincing." He looked up to see Parker extending her arm. A bandage wrapped her left wrist and forearm which lay supported in an elbow sling. Carl had a pressure bandage over his right eyebrow and appeared to be standing with his weight unevenly favoring his left leg. "Oh, I had no idea. Mr. Chu said you were stuck in traffic. Are you both all right?" He reached tentatively to shake Parker's hand.

"A little worse for wear." She took his hand, and shaking, noticed the restrained body in the corner. "Is that…Peck?"

"Odie Peck. I didn't know you actually knew him." Abadon still had her hand. He released it and perfunctorily took Carl's handshake, while looking down at the still-quiet Peck.

Carl's softly muttered "Son of a bitch" was drowned out by Parker's much louder "You bastard!"

She moved to kick the slumbering investigator but was restrained by both Carl and Chu. Grieg smiled from a seat at the small table. "Don't you worry, Parker. He won't go anywhere. Got 'im covered." He waggled his stun gun with a jaunty "how ya doing" wave.

Abadon, looking cramped now, edged behind Grieg. He addressed Parker. "I see you are very well acquainted with Mr. Peck."

Parker's glare of unbridled anger flared as she fired back, "THAT piece of crap tried to kill us! Multiple times. I would add today's attempted vehicular homicide to the Paris bombing and a train wreck."

Abadon raised his hands, begging off any responsibility. "I fired him. He is supposed to be out of my life and out of yours."

"What about your brother Richard?" She was flushed, pink at the temples. "Has he also fired Peck?"

"Yes, yes. I just got off the phone with Richard. Neither of us has had any contact with him since Monday. We've paid him off, terminated his research, and released him of all duties."

Parker took a full-bodied breath and sighed. She took another and blew it out through pursed lips. "So, you don't have any reason to have expected him to come here? Or to have prevented us from coming here?"

Seated now at the table, Chu said, "That man full of hate and mischief. I don't think that man can be just let go."

Abadon said, "No, maybe not. The authorities perhaps. You have my apologies for having hired him in the first place. Like I said earlier, I've had few dealings with him, and I, at least, never gave him authority to do bodily harm."

Peck groaned. Stirred. Everyone paused to see if the unwelcome visitor was going to rouse. He only groaned again in unconscious struggle against his bindings. Flexing his legs, he only pulled himself further into a fetal position, and relaxed.

Chu asked, "How did he know to come here? I'd understand if he followed you, but he arrived first."

Carl answered, "There must be a new tracker on our cars." He reached into a pocket to pull out his phone. "Or phones. We've already removed two of them. If they can store history, like mine can, he would have seen at least one of our trips into the mountains. Seeing that track to this place as the only one matching our recent trip up the I-5, he probably figured out this was our destination today." He grunted a short laugh. "Clever bastard."

"Too clever," Grieg said.

Andrew Abadon turned to look Carl up and down. The partner that was missing from yesterday's tête-à-tête, and former Abadon Industries sub-contractor. "Sorry we didn't get to meet yesterday; Mr. Reyes, is it?"

"Good to finally meet you, sir. You know, I did some minor touch up work on your servers, five or six years ago. Not exactly sure when."

"MacCauley says it was five years ago."

"Mac! How is he? Good man you've got in charge of that department. Hope you haven't been having any problems."

Andrew shot a glance at Parker, who was suppressing a smile with pursed lips. "Yes, well, maybe we can discuss that at some future time. But I really must go." He looked down at the still unconscious Peck. "What about that?" He gestured at the thin line of smoke wafting off Peck's shirt sleeve.

Grieg spoke up. "We really can't turn him loose. I'll go prepare a cocktail for him."

"You're not going to kill him, are you?" Abadon looked horrified.

Parker eyed him, wondering if she should go on the attack yet. If the deaths he had caused by policy initiative or disinvestments could have been tallied, he would rank among the world's worst killers. But policies don't leave bloody fingerprints, and disinvestments are simple business plan paradigm shifts. Clearly, he had never seen a murder and didn't want to.

Knowing glances were exchanged in answer to Abadon's question. Grieg turned back to him as he entered the small workspace in a side room. "No, not directly."

Abadon took a seat at the small dinette / worktable, relieved that the Peck issue was going to be taken care of and that he didn't need to know how. Parker and Carl, with simple gestures, indicated they wanted to join him, and Chu stood clear, waving a hand at the two empty seats across from Abadon, as a waiter might point out an available table. As she took her seat, she noted Peck's Beretta and pushed it to the side.

Abadon waited for her full attention. "Ms. Parrish. Excuse me, Parker. But I really must be leaving. I'm overdue elsewhere already. Your staging here is very impressive. It should be a museum piece, only I'm not sure what kind. Maybe a Disney diorama. Really a great production."

Parker's look of confusion resolved into understanding. She asked, "You think this is a production for your benefit?"

Carl added, "Really? First thing comes to my mind is 'you're shittin' me?'" His face was a mask of disbelief. "Chu? Are you a cast member in this production?" He lifted his chin toward the side room where Grieg had gone to mix Peck's cocktail. "Grieg, you orchestrate this production? You got an ASCAP card?"

Grieg called out, "Nah man. Don' know what an ass cap is. Don' know what an 'ass cap' card is. Don' think I want one."

Parker tried to modulate her voice to something resembling calm. "Andrew? I'm not sure what to say. Have you seen Grieg's presentation?"

"Cha man, we showed it." Grieg's irrepressible humor tinged his response.

Andrew said, "Yes, they showed me a power point, or something maybe a little more elaborate, a video." He raised his hands, open to the room, but inclusive of Southern California. Faltered, trying to frame the next sentence. "Look, Parker? This is Hollywood! MGM, ILM, Disney, Universal, hell, any number of companies across the mountain could have put that together in a few weeks' time."

"Chu?" Parker asked, "Can you crank up your diagnostics computer?"

The tall Asian looked puzzled at first, then got it. "Cha." He turned and wagged a few keystrokes in the air over a keyboard that materialized over one of the white rectangles on the work counter opposite their table when Chu's flying fingers expressed his passcode. The near vertical viewscreen's holo appeared, and in three seconds had solidified to opacity and began to display a rotating cylinder of topic headings. "Sorry, it's a lot slower than it was when this station was new." He chortled. "The comp I worked on back in Wellington was a lot faster. And it was a knock-off."

Abadon watched the display boot up. "Very nice. I've seen some development work on that sort of thing. There's a firm up in Palo Alto working on clear screen displays."

"Clear screen?" Carl reached across and ran his hand through the holo projection. "Clear screen? What about no screen?" He leaned forward, intently staring into Andrew Abadon's eyes. "I've seen a server room down in the valley that holds more information than the twentieth century ever produced. Probably as much as the first half of the twenty first, and it's powered by a still-functioning fission reactor the size of a dorm room refrigerator."

Abadon was unfazed. "Yes, I see, easy now." He leaned back slightly in response to Carl's intensity.

Carl turned back to the doorway; its shimmer had died. Chu had shut the physical door against further interruption. Carl then looked up at the dual digital clocks. "There. See those two clocks?"

"Yes, I noted them earlier. Easy set piece. I'd almost have expected it. But I would have expected a lot of dials, maybe some pipes and plumbing or loops of copper wire."

"Oh, there's plenty of copper wire, my friend," Chu spoke up. "The station is wrapped in copper cloth encased in a carbon fiber shell. Without an external antenna, this is a perfect isolation chamber. Nothing gets through but the occasional solar neutron. No bigee."

"What—?" Abadon stood ready to leave; he had expected a sales pitch but all this—.

"Dju ever hear of a Faraday cage?" Chu asked.

"Yes, I think so. Insurance against a solar flare?"

"There you are. Literally, here you are. We're in one big momma Faraday cage."

"What did you think about the door? The fuzzy electro-pop door?" Parker asked. "Didn't that trigger some sense of unreality? Make you ask yourself, Huh? Maybe just a little bit, a little WTF moment?"

"Well, I didn't think about it too much. A trick of polarized light or whatever. It's been years since I took college physics."

Carl asked, "What about the Last Day? Grieg, did you show him the last day sequence?"

"Cha, he saw it."

"Yes, they showed it to me." Abadon on the defensive, remained genteel. "Expertly done. Great effects. Or do your people call them F X?" Abadon looked pleased with himself for knowing a forty-year-old jargon term. Abadon grew nervously antsy, sat again at the table. Chu sat opposite and leaned across, his tone now intimate.

Chu said softly in tones that could not be mistaken for sincerity. "Mr. Abadon, two days after that sequence was recorded and transmitted, the three inhabitants of the Seoul Atmospheric Science Station got tired of watching the fires burning below, suited up, stepped out of their air lock, and pushed off for a final EVA." He shook his head at Abadon, as if unsure of either his stupidity or an astounding stubbornness. "As we later learned, the transmission was also received throughout central Africa, and across South America, but those societies had turned in on themselves after the obliteration of the First World and its life-giving exports. They degenerated into warlord-led tribalism."

Chu said, "Only a few places were left on the planet that hadn't been incinerated and had escaped EMP damage. Think of billions of watts of multi-spectral energy washing over the planet, frying advanced electronics and any claim to post-iron age civilization." Abadon started to object, but Chu pressed on. "The survivors? Yeah lots of survivors here, there. When the bombs stopped and the fires cooled off, survivors of the world's greatest cataclysm were slaughtered by the millions by rifles, then hand guns, then machetes as resources ran out." Chu grimaced. "Humans don't do very good under pressure."

Abadon pushed away from the table and stood. "Very good. I get it. If we don't change our ways, we're all going to hell in a boat. A convincing show, to be sure. Well done." His mouth warped between a thin-lipped smile and a grimace. "Parker, I am certain there are opportunities within your 'various endeavors' that we," he tilted his head slightly to the right, "my brother and I would find attractive." He waved an open hand, encompassing the entirety of the small room. "But whatever magic of perspective that you've pulled off here, to fit this room into the shell of a country house—" He paused, unsure. "But really, the end of the world video? I've seen as good in B movies. The city graphics are first rate." His nod of appreciation was belied by a smirk. He puffed a short heh, eh. "Really I do appreciate all the effort." A master communicator, his message was clear. He was done.

Chu spoke first in the thin silence that followed. "Effort?" His face was torn in real disbelief in contrast to Abadon's condescending show of doubt. Then louder, carotids pulsing. "Effort? You still think this is some Hollywood-styled production?" He began to step closer to his billionaire guest. Carl, now standing now in the crowded room, put a hand out to gently restrain Chu.

38

Abadon began to move toward the door, looking for whatever control opened it. Chu slipped from Carl's grip and moved to the doorway in a blocking move. Abadon frowned. "Mr. Chu, I've forgotten your first name. I'm sorry, but I really must be leaving."

"I'm Solomon, Solomon Chu," the smile was beneficent, charming, "but I've always been just Chu here. Mr. Abadon, if you will allow us one liberty, I'd very much like to show you something. It won't take more than a few minutes. I'm sure it will change your understanding of what this place is."

"Now see here—"

"He's right." Carl cut off Abadon's protest. "I've been out there; it absolutely motivated me to do everything I've done." Abadon wondered if that smile alluded to the theft of his millions. But Carl was insistent. "You *need* to take a brief look out the back door."

Parker offered, "It won't hurt, if you don't breathe in. But I'd recommend a re-breather mask."

Carl said, "Absolutely safe if they get the fit of the mask right. But they are right, Mr. Abadon. I think it will change your life."

Grieg had stepped into the tiny changing room and reappeared with a mask. To Andrew Abadon, its military no-nonsense design had some familiar elements, but it didn't look like the models his subsidiaries were producing for SWAT, DELTA, and SEAL units. His companies produced gas masks with rebreathers, night vision capability, and built-in com sets, but those were hodgepodge add-ons compared to the sleek, integrated design of the object in Grieg's hand.

His innate curiosity in technology took over. He accepted the offered mask and examined its interior. The lenses were opaque; the breather was actually a mouthpiece resembling a diver's mouth grip. He saw that, when in place, it would extend below the chin line halfway down his chest. Grieg reached out and flicked the on-off actuator. The screens lit, in orange shades, outlining the floor, his own and Grieg's feet nearby in simple wireframe.

"Just a moment." Grieg tapped another control below the mask's right ear. The displayed image went to a full color image as if the screens were completely clear.

Abadon looked up in appreciation of the technical advances over anything his companies produced. "This is really pretty good."

"Also bleeding necessary out there." Grieg indicated the door behind him with a slight jerk of his head. "I'll go first to make sure there's no one camping on our doorstep."

Parker said, "I'll go with you, Andrew, I've said I'd never do it again. To tell you the truth, I'm a little afraid to go out there again, but I won't ask you to do anything I wouldn't do myself."

Abadon stepped through the outer door to the air lock, to be followed by Parker. Grieg, identifiable by the Hawaiian print shirt he had been wearing earlier, faced them from a corner of the small landing. A sand-scoured safety rail saved him from a vertigo-induced pitch off the platform. He looked up from the ground below to a grim horizon line. A low cloud deck obscured the mountain ridge line across the valley to the north. Seeing that, Grieg said, "Great visibility today. We usually can't see much more than the slopes before us, and nothing of the valley floor."

There was no sunlit valley, no helicopter. Abadon spun around; there was no farmhouse. Taking stock, he noted the small metal platform was about twenty feet above a barren, eroded hillside. Bare indication of a two-rut track mirrored his memory of the dirt road his chopper should have been parked on. Even with cloud cover, the scene was completely wrong. He had arrived in late morning. Only a few hours had passed, if that, and it was clearly very near nightfall. He turned again to look out across the valley. A stiff breeze whipped at his pants, almost convincing him at last that he was in a different now. The rugged elevated structure behind him had more in common with his expectations of a NASA Mars habitat than a 19th-century farmhouse. He peered around the corner of the time port. A rusting lump surrounded by high brush were the remains of Tom Corliss' car. A small, translucent-skinned structure housed two three-wheeled vehicles. They appeared to be a blend of golf cart and ATV, but with an oversized traction tire in the back.

Andrew Abadon turned around again to look out across the valley floor. Neenach, California, in his early 21st century, was a homely, overlooked, agrarian outpost supported only by irrigation and hard work. The wreckage of the cityscape visible now surpassed his understanding. Mid-rise residential blocks were connected to taller office towers by fairy-wing sky bridges, all in advanced stages of decay from fire and weather. At length, he uttered softly, "Oh my God." Its mechanical echo in his earpiece startled him. His speech was garbled by the mouthpiece. He found he could disengage the tooth grips and speak almost normally.

Another voice in his earpiece, mechanical but obviously Parker, said, "God had nothing to do with this."

A third voice, Grieg's and just above a whisper, said, "I'm sorry, really no way to warn you."

Andrew Abadon turned to Parker. Her mask was identical to his, but he could see that she was staring out across the ruined valley. He rested his hand on her shoulder. When she turned to face him, he noted that a neat stenciled label over her eyepieces said Antonides. He asked, "You've never stepped out here, seen this?"

Her mechanically rendered voice responded, "Yes, once, at least a year ago, that was enough. Hate to admit I was a bit afraid to do it again, I'd said I never would." She added after a breath, "Carl's been down to the valley on some of the recovery missions."

"The what?"

Grieg's mechanical voice answered. "We go down there, into the valley. There are two cloud storage data sites down there. One is still cooking on its independent power supply; the other had to be rigged to a solar panel. We're mining technical data from before the Last Day." No one spoke. He added into the silence, "That's why we came here in the first place."

Abadon's skeptic's mode kicked in again. He remembered the mask's lenses were opaque when powered off. "How do I know that this isn't really good virtual reality gear?"

"Seriously? Still?" Her incredulity was evident, even over the mechanical voice of the electronic gear.

"Simple." Grieg answered. "Take off the mask, but only for a moment and don't breathe in."

Abadon considered; the view before him was a better 3-D rendition than any seen in the most advanced VR equipment they were investing in. The doorway. How had they created the appearance of an energy field? Grieg's ray gun or zap gun, or was it a laser? Or both? He put his thumb to the edge of the neck seal and hesitated.

Before he pulled on the seal, he turned to Grieg. "It won't kill me? It isn't toxic?"

"Oh, it will kill you sure. But not in a few seconds, and better you don't breathe in." Grieg nodded, thinking about it. "Yeah, sure. You can look without the mask for just a moment. I'll make sure you get it back on." He turned to face the taller man. "There's a sport now; be quick about it and don't drop it down there." He gave a quick nod to his right, indicating the darkening ground below them. "Take a deep breath and very slowly exhale." No one moved. "Whenever you're ready."

Abadon took several deep, purging breaths and finished with a chest-swelling maximum inhalation. He lifted the mask with two thumbs under the chin seal. He immediately blinked against some acrid agent in the air. Clearing the head straps, he pulled the mask aside and looked out across the valley. The sun had just emerged to his left, imparting a ruddy tint and hard shadows to the scene. Here and there, scattered among the towers, a few window shards or panes reflected the sun back. The wind from the west was hot, laced with fine dust. He stared, willing himself to disbelieve the scene, to come up with any explanation for the death scene in front of him. Neenach had developed from crop-circles and vegetable patches to a post-industrial mega-city and was then destroyed some fifty years before the present; if he remembered the presentation right.

He blinked against the wind, needed to inhale. Grieg was already reaching to help him back into the mask. The urge to let his last breath go and inhale was overpowering. When the last bit of rubber was folded over his chin, he took a breath. It was awful; it smelled of sulfur and soot. Ambient air had remained in the mask. He exhaled fully and the next breath was better, the third almost normal.

"Shyte man!" Grieg patted him familiarly on the shoulder. "You one brave Mama Slamma!"

"Andrew? Are you OK?"

Abadon remembered Parker was standing behind him. He turned to face her. "Yes," He suppressed a cough. "I wouldn't recommend doing that." He then coughed repeatedly and uncontrollably into his mask's mouthpiece. He nodded, imparting, he hoped, some sense that he understood they were standing in an alternate reality, then turned to look out again across the ruined valley.

As they viewed the wreckage of the Neenach City corporate complex, a dirt devil swirled up, eddying away from one of the taller vertical wrecks. It spun off to the east, lithe and beautiful, an out-of-place, transitory element of grace in a scene of unimaginable loss.

"There's a huge city here," Abadon murmured.

"Was," the pause was pregnant, but she finished it, "a huge city here."

"How many, do you suppose — died here?"

"No records to research; well, there are, down there." She pointed into the darkening city. "That's not why Chu, Grieg, and two others came here, to look for census records. To answer your question, I don't know, maybe a quarter to half a million. It would have been easier to have been on the other side of the mountain."

"Really?" His response was almost childlike, not comprehending the impacts of the multiple nuclear explosions on the western slopes.

"Instantaneous incineration would be preferable to seeing the mushroom clouds rising and waiting for the blast." Her voice wavered, even though the electronics. "Or surviving that, waiting for the rains. Or the mobs. I'd probably have killed myself."

The three masked forms on the porch, two invaders from the past and one refugee from the present, looked for a while longer in silence. Grieg was the only one looking below them for moving forms in the growing shadows nearby.

Abadon started, "How did you —" then stopped, still uncertain of all the implications. "How did you understand that this thing is what it is?"

Standing at the railing beside him, she was silent, staring into oblivion. "I knew it immediately, because they took me back a day.

A day I had to relive, in a surreal *déjà vu* with witnesses. This *is* a time station. They only told me when they thought the memory-cleaning drugs would make me forget. Apparently, I'm immune to that drug. This is a time machine. Our future invented time travel before it blew itself up."

His mind was turning with possibilities; she was speaking of the future in past tense. Odd that the recognition of that should cause vertigo. He grabbed the rail harder. "Can it go backward and forward?"

"No." Grieg's voice joined the conversation. They both turned to look at his masked face. "It can only slip back a little over a full day, and then back again. The verniers are screwed down to keep us from screwing up, except for the hour dial. We don't want to mess with them because, well, because we don't know how it works. What you used to say, it isn't 'mission critical' for us to know how it works."

Abadon looked out again toward the dead city, fascinated. "Do you go out there?" He remembered, "Oh! Right, that's why you said you're here?"

Grieg answered. "Yes, but we can talk about that inside. We should be getting in. We're safe up here, but down there," he pointed toward the row of stones that remained of the farmhouse's foundation, "there are going to be dangerous things moving around after dark."

~ ~ ~

They re-entered the station, Grieg insisting to Chu that there was no reason for real decon scrubbing because there had been so little blowing dust. "No, Chu, I'm telling you, it was a great weather day out there."

"You should still scrub."

Abadon asked, "That was great weather?"

Grieg ignored the comment and addressed Chu. "They don't have a change of clothes. We, they'll be fine. Don't worry."

"Decon?" Abadon said untangling his neck from the mask's straps. "We should be in decontamination? Why? What did you expose me too?"

"Air." Chu said from the table in the next room. "The air is still toxic, but if you kept the mask on, you'll be fine."

339

Grieg added, "He didn't take it off for long. Long enough to understand." He looked up into Abadon's worried face. "Just take a long hot shower when you get home. Maybe throw away those clothes."

"Speaking of home, I need to be leaving soon." Abadon stopped, not sure if his mouth was going to go on automatic. Something was definitely different now. He looked back toward the air lock door. He started to speak, his mouth hanging slightly open with the start of another standard closing. This *was* different. His clothing carried a residual scent of something acrid. "Parker, somehow I ought to resent your calculated manipulation of this afternoon. But that was sobering. I'm not sure how to react to what I've just seen. What you all have shown me—" After a pause, "Who would believe me if I were to share any of this."

"Don't, Andrew." Parker was now less than a foot away from him, looking up into his face. "I've been holding this in for almost two years now. I've only been able to make vague descriptions of what I've seen as if it was a vision from a nightmare. I've taken baby steps toward looking for an alternative to, to that." She glanced over his shoulder to the air lock. "What Carl and I have been doing since we first came here has been to methodically effect enough change that it doesn't have to end that way. With so much carnage, so much suffering, so much loss."

"How can you possibly do that from, what, over a hundred years ahead of the fact?"

"Change public opinion, on a wide variety of fronts; leak some of this technology, maybe a few decades earlier than it would eventually get discovered anyway. To jump start the tech solutions to widespread hunger and privation."

"I've spent my life building, creating, providing jobs, I—"

"You've spent a lot of that life funding phony think tanks, creating experts who will sift the data for the outliers while ignoring the median, the truth, the central truth."

"Truth?" He scoffed at her naivete. "Truth is what people want to believe. There is no absolute truth."

Carl, quietly listening, interrupted. "Andrew? Do you need to step out the back door again and reconsider if there is an absolute truth?" He stood from his seat at the small table. "Sure, you can go out the front door, hop in your chopper and catch your jet stream back to Kansas. Will you ever be able to forget the truth lying outside the back door?"

Abadon's posture changed perceptibly: his arms were tucked in at the waist, hands balled into loose fists, his head lower on his shoulders. He seemed to have shrunken a size in every dimension.

Carl came forward with outstretched hand and offered a small, dark cube to Abadon, who took it and turned it over and over in his hand. "What's this?"

"A few hundred of whatever comes after 1,000 terabytes. Its skin charges the unit from ambient light; it transfers data through its base when set down on a data port. Early versions of this data cube are several decades out, if we wait for it to happen on its own."

"Take it, as a memento. I've already taken off all the data." Grieg said.

"That's not even world-changing tech, Andrew. We emailed the formulation tech for laminar carbon batteries to car companies. We could leak the chip design to labs, but that's not the kind of tech that's going to save us from ourselves. A few of our beneficiaries are already setting up research departments, with grants from Room for Tomorrow. We are investing in wind, and tidal energy products." Brows furrowed, she thought of the list and how his fortune could help. Their conversation in her boardroom came back to her, the second thing.

As if reading her mind, he asked. "What do you want from me?" Andrew Abadon, ever the calculating corporate superstar, had drilled down to the pertinent question. "What can I do? Yesterday, you said there were two things I could do. I'm onboard for the first. This trip was the second, right? Is that it?"

39

Parker answered immediately; she had been formulating the answer for days. "Stop getting in the way." When he started to object, she shushed him.

"Change your own outlook and then change the world's outlook. You fund too many politically expedient organizations you don't even believe in because they support your corporate interests. STOP it! Christ, Andrew." She took a deep breath, "It's all complicated, but at its core there is one serious flaw with unchecked global capitalism." She blew it out. "It depends on growth. There is no current economic model that considers success to be sustainability."

"Your quarterly reports would freak out your investor pool if you predicted zero growth over the coming four quarters. But what if you promised them a steady supply of dividends? What if all you expected of your existing companies was that they simply paid dividends on profits? I can guarantee you profits. That's what sustainability will get you: a long line of profits based on ensuring a supply of needed energy and commodities. You need to change your model from perpetual growth, or you will continue to drag our world economy into an endless cycle of grow and bust, grow and bust." She spread her hands open wide. "In the long run, no one wins."

She leaned back in her chair, staring at the person who had been near the top of her adversaries' list. She had his attention. "What would happen to the modern capitalist movement if you, Warren Buffett, Bill Nye, and Robert Reich were all on the same page? If the four of you took a world podium tour to preach a new world order?"

At their feet, Odie Peck shifted, another of the small moves he had been making for the twenty minutes he'd been conscious.

Bullshit, bullshit, bullshit. These powerplay assholes were discussing the future as if they could do something about it.

He groaned again, just enough to be heard over the arguments above him. He curled again as if some internal spasm was forcing him into a fetal curl. Almost, almost. He reached for the edge of his boot with a tentative finger. Damn, still too far from the grip. He relaxed again and allowed the drugs the little Russian prick had given him to provide the rest he needed.

~ ~ ~

A quick phone call in the farmhouse hallway sent the chopper for three large pizzas and a few quarts of diet soda, and they settled in to discuss what really could be done. Tomorrow. Next week. Next year. The conversation at the table was on par with a sequestered meeting between Russian and American diplomats in a quiet corner of the UN. Policy, strategy, cause, effect. No one was convinced that any one act, or any small collection of acts, would stave off a thermonuclear exchange. All agreed that a concerted effort to affect an "all of the above" strategy would at best slow global warming. That was, they concluded, the land-robbing trigger that set off the Last Day.

New technologies were going to be needed and would need to be fed to some of the best R&D labs to be implemented twenty or fifty years ahead of the present development curve. Some new techs, including using Bucky fibers for electrical transmission, doped carbon layer batteries, and increased step-down applications for small appliances, could be brought online relatively quickly. The small fusion core reactors that would become ubiquitous in the 2120s were more difficult but could still be in use decades sooner. Increased wind, ocean current, and solar energy would bridge the gap. Even extending the life of existing fission plants was preferable to coal.

The coal industry was going to be taught that its product was more valuable converted to storage and transmission materials rather than simple BTUs. The petrochemical industry also needed to learn to glean the last possible therm from its gasses and to slow production even further. The remaining in-place fossil fuel was far more valuable as plastic and other by-products than as automobile fuel. Abadon balked a number of times. And they had almost forgotten about the intruder in the corner.

But global warming wasn't the only issue, the only trigger. Education could create an even playing field. Plans were laid for satellite time in the very near future, for beamed broadcasts into oppressive and restrictive regimes. Broadcasts that would emphasize, sometimes subtly, and sometimes bluntly through different campaigns: the importance of birth control, and ubiquitous literacy, and math skills. Only education could fight extremism. The lessons of Goebbels and Breitbart were taken to heart. Fake news had to be countered.

And beyond a touch-up injection to keep him sedated, they had almost forgotten about the intruder at their feet.

~ ~ ~

When he had gone, more than six hours after his arrival, no one was absolutely certain that the attempted conversion of Andrew Abadon was successful. Instead of going over it in great detail they ate a simple meal of leftover pizza, salad, and veggie sticks with hummus. Crunching a celery stick, Carl asked the room the obvious question on everyone's mind. "Do you think he bought it? That we did any good?"

Chu answered, "My guess is that he'll contact someone from his tech companies and get that cube analyzed. He may have some trouble explaining where it came from. The photovoltaic shrink fabric alone will revolutionize outdoor equipment and outer wear. The chip construction inside is several decades ahead of Moore's law. And the interface, hell, a good designer will be able to figure out how to start building those next year. You said there are kid games that do that already?"

"Yeah, my nephew in Florida has one," Parker said. "That's only token tech though. I wanted to steer him closer to battery storage." She thought about the meeting, what had been missed. What had been said. "Well, if we've done anything, what we really need to happen is an incremental paradigm shift in climate change acceptance by the forces of no."

"Shyte."

Everyone turned to look at Grieg. "We never even told him about the fusion plant that's keeping this place running. Or the advanced photovoltaics powering up our go-cart."

"So-K," Chu said, "don't worry about that. They won't learn how to efficiently crack uranium for technetium ninety-eight for another fifty years, ten more before they bind it to chromium. We have plenty of data here." He turned and patted the console behind him. "And we should probably get a lot of it offloaded to the present just in case."

Carl asked. "Onto what? What have we got that can hold that data?"

Chu pointed to Peck's slumped form on the floor. "Get that peckerhead outta here tonight. Come back tomorrow with all the mass storage you can find at office supply. We'll burn up a USB getting it down."

"You sound like you're in a hurry; what's the hurry? You've been here almost two years."

"Hurry, shyte yeah we need to hurry. Damned future is going to get unstable."

Parker and Carl exchanged a glance then turned back to Chu. Neither asked.

"What if you did change that man's mind? What if he does what you want? What if Room for Tomorrow becomes the, what do you guys say, the game changer?" Neither answered, baffled. Chu raised outstretched arms, panned his head to take in their surroundings. "The future changes! You come back next week, you may find someone here from the other future, another future, doing graduate paper research on the Downfall of the American Super Culture in the Early 21st Century." He indicated one of the consoles with a nod. "That's what they were working on, when their world ended."

Parker said, "Or we might find a crumbling 19th Century farmhouse. No frills. No time machine."

Grieg muttered a simple "Shyte."

~ ~ ~

"Uhhnnn." They turned and watched as Peck tried to roll over. His bound legs' attachment to the table leg prevented all but the simplest movement beyond stretching or coiling back into a fetal curl.

His groan brought their conversation back to the present. Parker watched Peck as he strained against the restraints without a shred of pity. His eyes were still closed, but that did not mean the man was still unconscious.

She reached down, dug a fingernail into the man's eye socket, and pushed.

"HEY!" Peck struggled to move his head away.

"Welcome back. Enjoy your nap? Peck, Pecker, Peckerhead. That's what they called you when you were a little too small on the playground wasn't it? Peckerhead?"

"Bitch!" He'd been taunted with the obvious nickname through grade school, but only one man had tried it in the Navy. He tried again to rise, but his wrists and ankles were bound in an unyielding but flexible material he'd never seen before.

She poked with the fingernail again, this time under his chin in the soft flesh behind the bone. "Murderer." His Beretta still lay on the table. She grabbed it by the barrel and saw the red dot of the safety indicator. "You simple idiot. You left the safety on!" She tapped him hard on the knees in malice. The pistol made a good hammer.

"Oww!" the complaint was heartfelt, but contrition was a long way off. "You can't prove shit. I want a lawyer."

"I am a lawyer. What do you want to discuss: your email activity to accounts in Italy immediately before each bombing, chain of evidence, tracer molecules in batches of HMX? Chemically conclusive lab results linked the Paris bombing to the Croatian train bombing. You, you son of a bitch, have all the lawyer you're going to get."

"Parker." Chu tapped her on the shoulder. She looked up to see his hand held a syringe full of a translucent yellow liquid.

Peck saw the syringe too. A trace of fear gave way to defiance. "I blew your house, assholes."

"What'd you say?" She rapped hard on his bound knees again with butt of the gun.

He grinned through gritted teeth and flinched again, but only slightly. He'd seen this one coming. He maintained eye contact but relaxed his posture, slumping and sliding his hands toward his ankles, closer. "I think when you get home, you're going to love your beach house." He laughed, coughing. He stopped when she whacked his shoulder hard with the gun. "Your house." A cough. "It's at the bottom of the cliff. I blew the pylons after you left the house today. You were supposed to be in it, but you left too soon."

"Jeezus, you are a fuck up." She raised the gun up to hit him again, but Carl restrained her hand, and unsuccessfully tried to remove the gun.

"Park, Parker, honey. Are we going to add torture to our resume?" Carl's voice was reassuringly calm in an ocean of violent emotions

"No. No, probably don't want to go there." She turned back to face Peck's malevolent stare. "You idiot, those pylons only hold up the pool deck and pool. That's fixable."

"You think you're so damn smart, Ms. Smart-ass lawyer lady from back east." His voice had lowered to a venomous whisper. "The pylons, the kitchen, the garage with your fancy cars. I heard the separate explosions. You're the proud owner of rubble."

Parker dug the nail of her middle finger hard up into the soft flesh of Peck's jaw. "You're lying; the alarm would have sounded if you were in the house."

"Not in the house, bitch. One shape charge aimed at the garage, the other at your bedrooms. The house would have imploded and collapsed. I didn't get to watch, but I heard the blasts and the dust shower was still falling when I left the neighborhood. The place is probably lousy now with forensic cops trying to figure out what the hell happened." Through the pain in his jaw, his grin was pure malevolence.

Parker leaned back against the counter, fist over mouth. *The house was gone. Jazzman!* She'd shifted her hold on the gun from barrel to grip. She snicked off the safety and lowered her aim to his knee, ready to pull the trigger, her face contorted in anger. "Goddamn you son-of-a-bitch." Peck seemed to cower, curling with hunched shoulders, his cuffed hands pulled down to his ankles, his head turned away from any blows. Carl stood, raising Parker's gun hand, and stepped between her and Peck.

"I've heard enough of this shyte." Grieg stooped down with the syringe at hand.

Peck cringed, pulling into a smaller ball. His bound hands came up with a small chromed Colt revolver. His first shot hit

Grieg in the left shoulder. As Grieg collapsed on Peck, the syringe's needle jabbed Peck's shoulder, its chemical load intact. Peck squirmed to shift Grieg's weight off and searched for another target. Parker was a mute statue, hand over mouth in horror. Carl was moving.

Carl dove into the crowded corner, blocking the pistol and reached for the syringe's plunger. Peck squirmed again, barely aware that Carl had just delivered the syringe's opiate cocktail and fired twice. Carl's surprised "Oh!" was drowned out by the third and fatal shot. It was over in less than sixty seconds. Peck's body relaxed, sedated. Carl's body slipped off to the opposite wall and lay still.

Parker screamed.

40

Carl is dead! Parker leaned back against a wall, exhausted, emotionally spent.

It had all been so fast. She tried, struggling against the calming drugs Chu had provided, to piece together the final few seconds of Carl's life. Peck, that son-of-a-bitch, had a boot gun. Stupid of them not to have looked.

Carl is dead. The sweet boyish grin, gone. His subtle hand hold with entwined fingers; she didn't like holding hands that way, but somehow…

Carl is dead. Another wave of weeping shudders took her over. But she was dried out. She'd need an IV drip if she couldn't stop.

Chu had been great, helping sort out the heap of tumbled bodies, helping her do everything possible to stabilize Carl, and finally, reviving and patching Grieg. But she couldn't get beyond "then." There had been two shots, Carl had tried to lift himself away from Peck, and then there was the third shot. *Carl is dead!*

And Peck was dead. Chu had finished it. Taken the Beretta and put two rounds point blank into the man's heart. But it didn't matter anymore. Carl is dead.

~ ~ ~

By nightfall, Chu had taken charge and removed Peck's corpse, carried it out through the airlock, and dumped the body over the railing. He'd taken Grieg downstairs to the farmhouse kitchen and gelled and patched his injuries. The wound had been a through and through just below the collarbone. Then he'd sedated Parker. She was fitfully dozing in Noel's bunk. Previously, Chu had stripped Carl's shirt and with Parker's help had tried every intervention he could think of, but his training and the limits of his field medical pack were insufficient. Gel packs could only stop external bleeding, and the huge purple blotch that spread across Carl's abdomen presaged the ultimate result of the three gunshots.

He was at a loss. Carl was going to go missing in the 21st century. He was now a high-profile citizen, victim of two previous bombing attempts. He could only think of one person to contact in the 21st century that might have the resources to help. He found Parker's jacket and then her phone. Being careful not to disturb her sleep, he gently pressed her right thumb against the button and saw it come to life.

It picked up in two rings. He could hear a soft turbine whine in the background. "Hello, Mr. Abadon? This is Solomon Chu."

"Mister Chu?" His voice had the arc of surprise. "Is something wrong?"

"Yes, Sir. Somethings have gone terribly wrong. I don't know how to navigate the legal world in your century and I'm going to need—" He paused, flicked his glance between Parker and Carl. "We are going to need an extraordinary bit of help."

41

Parker slid to a gravel-slinging stop in front of the farmhouse. She had been driving too fast for the drizzle-wet, rutted track that passed for a road. The brakes on the new Model X were harder than those on the ruined Model 3. She pushed open the usually locked door to the kitchen and found a note and three data cubes on the kitchen table. She saw that the note was addressed Dear Parker and was handwritten in Grieg's difficult-to-read scrawl. She ignored it, knowing she could pick it up on the way out.

She had news and was anxious to share it. The cliff house had been saved; the claymores had blasted against the concrete walls but only blown away portions of the overhanging roofline. The house and Jazzman were safe. Only the pool deck had collapsed into the Pacific surf. More importantly, Abadon had made a huge splash in the news. He'd fired the science deniers in his PR firm and already been on O'Reilly's show, talking up his change of heart and mind.

She bolted up the stairs, taking two steps at the landings. With a heart-stopping pause, she halted at the top of the stairs. The side doors were shut, as they had been so many months ago. The hall was as dark and foreboding as it had been on the first rainy afternoon when she had come almost two years earlier to conquer the nightmare, to put an end to the night terrors. The only light in the hall filtered up from the small window halfway up the stairs.

She was suddenly overcome with a wave of nausea. She had to bring to mind the many things that had happened. Yes, they *had* happened. She was the CEO of the world's most innovative agency for social and environmental change. The love of her life should have been at their Anaheim offices, reviewing plans for undersea waveform generators to be placed beyond the La Jolla sea grass beds. Instead, his ashes would be interred in a sprawling East Los Angeles cemetery, next to his mother. It had all happened, dammit!

She fell against the wall as her knees began to buckle. Sweat beaded on her temples and forehead. She took a step and pushed the first door open to the left. No change. She took another and pushed the door to the right open. Same result. Neither of the two side bedrooms held more than dust and scraps of paper.

The crystal knob on the front bedroom door gleamed dully in the dim light from the two opened doors, but there it was plain enough. Ominous. The door was closed. The door of her nightmares had usually been open for most of two years, protected by either an energy shield or the inner physical door of the time chamber. Her pulse hammered in her temples; she could actually feel her heart banging in her chest. She put her hands over her heart, feeling the rhythm. She closed her eyes and took a deep cleansing breath, and another. Slowly her heart rate dropped. She found her hands were covering her face. She dropped them to her side, her fingertips moist with tears.

She opened her eyes again and closed the five steps to the end of the hall. She reached for the handle and grasped if firmly. It took more force of will than anything she had done recently to push on the door, but in the end, it swung open freely, revealing a sunlit view of the mountain sloping away to the east and, in the distance, the pitiful, checkerboard fields of Neenach. She stepped into a dusty, wood-paneled bedroom. Bits of paper and empty chip bags and snack wrappers littered the floor along with a few empty soda bottles.

Outside, beyond the grimy, ancient window glass, a light wind blew dust up on the valley floor, but she couldn't hear anything outside. She heard only her ragged breath in the quiet. "Oh my GOD!" She slumped against a wall and slid slowly down. Unsure if two years had passed, unsure if she might have hallucinated a highly elaborate alternative history or future. Panic robbed her next breath. Her fingernails cut arcs into her palms. Her lower lip caught between clenched teeth. She began to pull out of the panic attack when she tasted blood. She gulped and drew breath, and another, and more too fast. Hyperventilation began to make her dizzy. *Stop it!*

She willed her breathing to slow again, swallowed the bile that pushed toward the back of her tongue. She got to her hands and knees and slowly stood.

Validation lay just outside. She went to the north window and looked down. There was her own brand-new Tesla; she ran to the back bedroom window and looked down to find one very dirty derelict on blocks. Tom Corliss's car. "Oh, thank you, Christ, if you're listening." She rarely called on Christ anymore, but this was an exceptional salvation. She remembered the data cubes and Grieg's note, and fairly hurtled down the stairs to find them.

In Grieg's attempt at block printing: "Parker. Hoping you'll find us upstairs, but just in case... there've been some changes. I came back from Neenach last night, picked up the last of these cubes. One should be helpful in stopping the black rice blight of 2043. The other two, bits and pieces, you'll find interesting. But the big thing is we've been in a bit of a row with the Bigs in NZ. They insist that we're trespassing and have never heard of us. The timeline has shifted, we don't know how much. Leaving these in case the changes become too weird. Out here, they should be safe. I've got to go back up to talk with Chu, hope to talk with you this afternoon. Hugs, Grieg."

Parker grabbed up the three cubes, looking at the fine silver fractal filigree that so defined them as something not of this century, her 21st century. She folded the note on the same fold lines Grieg had made and walked slowly back to the car. If the timeline was changed, why wasn't there a research group in there researching the causes of the downfall of the great American Century? Where was the time portal that had been hi-jacked by the visitors from five decades after the Last Day? What had happened to Chu and Grieg? Was Noel buried in the back at the base of the stone wall? Or would that not have happened for another two centuries.

Did it matter if the future didn't invent time travel? Maybe they just picked another location. Too many maybes. It would be far better if the future didn't blow itself into radioactive oblivion? Parker found herself standing beside her replacement car. A new Tesla, Model X, black-tinted windows on custom RFT green metallic paint. Musk had it built for her in thanks for the upgrades to his battery tech. The new one had black epoxied wheels and her initials over the door handle in gold. Nice touch she thought. She also thought about her choices. Each day is a new beginning. She sighed and tossed the three data cubes with who knew how many petabytes of information stolen from a possible future into the passenger seat.

She turned to take a last look at the humble little house, a refuge for some past owner who thought he could make a living grazing sheep on these rocky slopes. Now a forgotten bit of the past, discovered only by those who found a way onto the closed pathways of the Los Padres National Forest preserve. It did look better with greening hillsides framing its clapboard siding. She recalled the visit with Carl, waiting in the drizzling rain for her other self to appear on that long ago yester-today. *Oh, Carl, I do miss you.*

Past tense, future tense. Preserve the past, protect the future. The future was an uncharted roadway of limitless possibilities. She shut her eyes, thinking of Carl, his preference for cremation would be honored. Her fists knuckled her eyes to stop another burst of tears.

She looked north, toward a future sky-scraper skyline and filled her lungs with the uncommonly clean afternoon air. Neenach was still a Mondrian patchwork painted in irrigated green and desiccated tans. The drizzle had not affected the tiny community. A dust devil dipped and danced across the valley floor. Reluctant to leave, for perhaps the last time, she finally got into the sculptured seat of the luxury electric roadster and touched the start button. She wore a grim smile through tears as she pulled onto the narrow track down to the valley floor.

Parker slowly turned a bend in the jagged mountainside road and saw two men approaching along the shoulder. One tall, bald as a boulder; the other, a head shorter, one arm in a sling, had tousled, sandy-blonde hair. The taller one stopped, cautious, not recognizing the car. The shorter one, saw her, hopped excitedly, and began to run toward her.

She vibrated with absolute joy as she jumped from her car. But as she felt Grieg's reassuring one-armed hug, she felt suddenly ill; a nauseating urge to purge heaved her diagram. *What?* A silvery shimmer began in her peripheral vision. She was aware that gravity no longer pressed her to the ground. Grieg's hug became insubstantial; he faded from view. Beyond any expectation, she found herself at the communal table in CalStation. Not the farmhouse, but the overcrowded time station with too many people in it.

~ ~ ~

"That's enough, it's over." Chu's tone had a finality she didn't recognize.

She wished she'd had a tape recorder running, so she could make sense of things later. Would that have made a difference? Grieg had started it, Chu had tried to slow him down, to make it more understandable, but the flow of time simply confused her. *But Carl is alive.* He was at his house uploading the most recent data cubes. Her house was a crime scene tape strung around its perimeter. The cantilevered deck and pool were rubble in the surf and rocks below. The floor of the sunken portion of her living room hung from damaged rebar from the remainder, and kitchen and garage walls were damaged but intact. *But Carl is alive.*

Peck had been bundled into Noel's unused bunk while they transmitted a massive data package to Wellington. *And Carl is alive! Why do I remember that he'd died.* She had been preparing to take the data cubes to the RFT headquarters to give them over to Brayton and survey the damage to her house. Yes, she knew about the house; how is Carl still alive? Right. The Gods of temporal paradox began to restore her memories. As soon as the data package had been sent, both Chu and Grieg had made a move toward the time verniers.

The suggestion to back up a few hours had bubbled out of them both in much the same way as their reporting of the events did now. The three of them, Chu, Grieg, and Parker, had determined exactly when to reset the verniers to alter the chain of events the previous day when her world crashed. After Abadon had gone and before the moment when Carl died. *But Carl is alive? Why don't I remember this time that I was there, then? I used to remember then. That then...*

But she *could* clearly remember now, in almost perfect color and smell sensory memory, the crucial moment. She shut her eyes to recall the moment of the flash of light. The moment of the re-entry into the time loop.

~ ~ ~

"Not in the house, bitch. I aimed one shaped charge at the garage, the other at your bedrooms. The house imploded and collapsed. I didn't get to watch, but I heard the blasts and the dust shower was still falling when I left the neighborhood. The place is probably lousy now with forensic cops trying to figure out what the hell happened." Through the pain in his jaw, his grin was pure malevolence.

The flash of light was bright, in fact, brilliant. Blown transformer bright, flashbulb bright. And the lower gut nausea of a time jump, but the hatred flaring in Parker drove through it. In that hatred, and in that moment, her peripheral vision sparkled in dancing motes of gold.

She settled against the counter, fist over mouth. The house wasn't gone in this pass through. Jazzman was safe. She'd shifted her hold on the gun from barrel to grip. She lowered the aim to his chest, ready to strike, her face contorted in anger. "Goddamn you son of a bitch." Peck seemed to cower, curling with hunched shoulders, his cuffed hands pulled down to his ankles. His head turned away from any blows. Carl stood, raising Parker's gun hand, and stepped between her and the intruder. Peck, already curled into a corner, began to shrivel, with his ankles bound to the table leg, his body pulled into a fetal curl.

"Enough of this shyte." Grieg stooped down, syringe at hand. He kicked Peck in the shoulder and then kneeled, placing his full weight on Peck's chest, flattening the gunman's back to the floor. Chu pounced in front of Parker and pulled Peck's reaching hands away from his boot. With his other hand, Chu retrieved a small silvered pistol from Peck's boot. He stood and tossed a snub-nosed revolver aside then turned to Parker, blocking her final shot.

"That's enough, it's over." Chu's tone had a finality she didn't recognize. He had been difficult, almost mean in his disclosures of how terrible life was in his world. She thought he was capable of emotions she'd only glimpsed. But this hard-edged command voice was sufficient. She lowered the gun to her side, her shoulders slumped. She raised the gun, grip first, and gave it to Chu.

"I want that bastard dead!" She was surprised to hear herself saying it with such vehemence, but it was out. "That son of a bitch tried to kill us in Paris, almost killed Carl in Croatia, earlier at my house, and just now. I think he would have happily killed us all."

Peck had gone limp; the drug had rendered him immobile, if not yet unconscious. Carl stood then helped Grieg up. They shook each other's hands in grim victory. Carl turned to Parker. "Sweetie, whoa. We don't have to kill him here we can dump him out the back."

"I — I." She stared down at the would-be killer as Peck's eyes slowly lost focus and closed. "Whatever plans you've made are over, you miserable — "

Carl put a hand on her shoulder. "It's OK. Like Chu said. It's over."

~ ~ ~

Plans were laid. It was almost time for the next satellite flyover, and data needed to be sent off to Wellington. Carl pulled Parker aside as Grieg and Chu began putting on gear to greet the 23rd century dawn and find their satellite. "Hon, I need to get to my house. I'm going to get into Peck's business. I need to show a few hundred thousand moving from the Abadon's A-Bank accounts in the Caymans into a Bank Suisse account for Peck. Then I need to book him a hotel in Hong Kong or Bangkok. His traces have to leave Kentucky and LA County."

"What about the real guy? The bastard." She cast a glance over Carl's shoulder at Peck's drugged form and the shrink ties securing him to the bunk.

Carl shrugged. "I don't care. The guys can take him out the back door and drop him off the platform for all I care." He went over to the limp body and began to pat it down for any useful information. He immediately found resistance inside Peck's jacket and pulled out a folded single sheet. Opening it, he whistled. "Forget Bangkok. He already has reservations for Grand Cayman Island. That will simplify things immediately." He showed her the familiar form of a home-printed Q-coded airline ticket.

She smiled back at him. "He was going there anyway!"

"Yeah, though I think he intended to finish us off first, then leave."

She frowned now, then closed her eyes in thought. "If Andrew was not lying to us. If, in fact, he and his brother both fired Peck." She looked down at their captive, her gaze hardening. "It was vendetta. He had missed killing us in Paris and Croatia, so he wanted us dead out of spite. Oh, yeah, and the house. He'd tried to blow us up at the house, but we left before he could blow it."

He put an arm out, resting it on her shoulder. "Hon, the house!"

She shot a look at Chu. He responded, shaking his head, no. "If we go that far back," Chu said, "we have to intercept him before he comes here, before Andrew gets here. We don't know if we'd get the same outcome without having to go through that initial scene again. We have to leave the loop jump where we did, just before he pulls the ankle gun."

She exchanged glances with Carl, who was nodding in agreement. "He's right, Park." He said. "Going further back there are more variables to get in front of. Right now, we know Andrew Abadon was convinced that this whole thing is not a Hollywood production. At least, that was the outcome with everything that will or would happen if we let history ride up until that moment."

Chu added. "We've already looped back once; you've gotten the result you needed from Abadon, and we've nailed this twit. I say we leave the loop back where it was. Who knows what else may have happened in another pass through?"

~ ~ ~

Carl felt faint. He sagged onto the narrow bench of the CalStation's communal table, shaky and unstable. Curling into the wall, his feet bumped up against Peck's bound and secured ankles. "Peck! That piece of shit tried three times to kill me."

He heard, as if from a distance, "Are you all right?" It was Parker. He felt her hand caress the back of his head and slide down to his back where it paused. "Carl?"

"I'm good." He opened his eyes, forcing focus. "It finally dawned on me. I died." He kicked at Peck's legs. "This bastard killed me, and I lost count of how many others in Paris and on the train. The other victims in the body bags." He shook his head; some of the sudden fog seemed to lift. "I need to get him out of here."

Sudden mobilization hit the survivors of the second-round pass through of the attempted murder. Peck was untied from his temporary storage on Noel's bed and dragged toward the air lock to the 23rd. As Carl watched, still weakened, Chu and Grieg prepped for a short excursion outside. Masks in lieu of the environment suits required for longer trips. Sooner than he'd have expected, they were back inside the decon chamber with the nitrox blast flushing the 23rd's toxic soup from the chamber. When they'd hung their masks on the proper hooks all was silent. The quiet hum of the station's nitrox generator and the persistent drip from the shower were the only sounds.

"He's gone?" Though Carl knew the answer, he didn't know what else to say.

Grieg said, a grim, thin-lipped smile playing, "He gone. Didn't even bother to untie him." He gently patted his shoulder bandages, then waved his good hand in small circles. "Whatever the dogs leave, we'll clean up later."

Chu said. "We need to make another trip down to Chick-a-Saw. There's a bio-chem's back-up I found last trip that I didn't have room for on the cubes. They were working on chromo-genesis solutions to radiation poisoning."

Parker asked. "Radiation poisoning can be cured?"

"Cha, it's hard, but so many of our people back in NZ could use this. We knew the therapies were developed. They helped nuclear plant workers who suffered overexposures and even some of the asteroid belt workers who took a cosmic radiation blast. Last trip down valley was the first time I even got close to finding it. The files were under the original university drives, not the pharma." He shrugged, "I got lucky. If we can't prevent the Last Day, NZ is going to need this."

With Peck's body no longer around, Carl felt his strength returning. The idea that he had been killed, or re-incarnated, or whatever took a back seat to the revised future, his revised future.

Carl put a hand on Parker's shoulder, for connection and for support. His strength was coming back but his body and especially his ribs were still sore from the train accident. "Babe, we need to get moving. They have work to do, and we do too. Why don't you take those new data cubes down to Brayton for downloading? I have to get back to my house." He rephrased. "I need to get back to my house and try to erase Odie's recent past, anything he did here in California in the past few days."

"Why?" Her expression pinched with refreshed anger. "Peck is gone, thrown out with the garbage."

"Why? I need to erase as many traces of Odie Peck's existence as I can since he left Topeka; recent purchases, rentals, etc. needed to be purged from the web. The man needed to disappear. His trip to California needs to disappear. I have to hope that he didn't leave fingerprints when he bombed your house." A nagging detail solidified. "And he has a car, somewhere." Thinking back down the timeline to before the loop back. "He followed us here in something."

Chu held up and rattled a key ring with a Chevy-branded car fob along with several other keys. "This was in his pants, pulled it before we dropped him over the rail." He paused, looking inward, as if pulling up a memory. "Carl, first time 'round, I called Andrew Abadon after everything went down. He agreed to purge any record of Peck from their records, not sure if that actually happened since our replay occurred before that call. He was open to that if you want to call him."

"Great." It was coming together. "Park, why don't you take our car back to my house, I'll take Peck's car to East LA and abandon it with its keys. If it makes it to sundown, I'll be amazed. It will either be reported or chopped, not our problem. But it is a problem if it's found near here. I'll Uber back to my house. See you there."

42

The evening air at Carl's house had been freshened by an offshore breeze. His house didn't offer the Pacific Ocean view that Parker's La Jolla cliffs house had provided, but it had high walls, palms, tropical landscaping, and a waterfall pool that encouraged relaxation. Parker was relaxed. One of her biggest problems could no longer make problems in the 21st or the 23rd. Odie Peck was dead and gone. A week had passed since the replay of his treacherous visit to the time station.

She and the new refugees from the 23rd were recapping the day and agreed that they could all share Carl's three-bedroom bungalow until new arrangements could be made. Specifically, Carl's contacts would create new ID kits for Chu and Grieg. He would insert their school records in appropriate immigrant-heavy cities. They'd also need forged birth certificates from the US, and finally, employment for them in the RFT business structure.

Grieg had just agreed to test into the IT department when all stiffened at the sound of the doorbell. Parker stood, signaling all to stay put. Carl pulled up the security app on his phone and called out, "Calm down guys! It's Brayton." From his view of the home's perimeter camera displayed on his phone, he could tell there were no other vehicles nearby other than her sleek new electric Mustang.

He stood and moved toward the front of the house. "I'll get the door; you all stay put." Parker heard his bare feet padding toward the front door. Her shoulders relaxed, dropping an inch as her flight response softened. She sat again, this time with a drink in hand. She raised it in a toast. "To a future when the doorbell is just a doorbell, signaling pizza, or friends, or a package."

Chu took a sip of his margarita and added. "To a future where I feel like I belong." His sidelong glance included Grieg in the next toast. "To a future where we feel like we finally belong." They all raised stemmed glasses or tumblers to their lips.

Grieg cleared his throat and was about to add a line when Carl returned with Brayton close behind. Joanie was animated, smiling eyes and a toothy grin. She pointed an index finger at her wristwatch. "Turn on the news. Probably any news. Try MSNBC or CNN." She saw that she was blocking the view of the large flat screen on the porch and took a seat on the couch beside Grieg.

The first images of a life insurance commercial were met with a chorus of groans and jeers from the lightly intoxicated group. Carl slid into an empty space beside Parker on a rattan couch. The porch's flat screen soon left the parade of commercials and went to Anderson Cooper giving a rundown of the latest minutia in changes in Washington political maneuvering. The news menu to his right indicated he would soon be getting to "Trouble in Topeka." The scroll line below ran a headline with promise: "Major policy shift at Abadon empire HQ has energy stocks in turmoil."

Brayton pointed out the info lead scrolling across the bottom of the screen. All attention now was on the flat screen. Carl boosted the sound slightly when the gray box highlighting the topic menu dropped down to the *Trouble in Topeka* line.

Cooper's familiar voice reading from the tele-prompter; "Out of Topeka this evening, we understand sibling rivalry at the top of the Abadon empire has Richard Abadon on the verge of separation from the financial empire on news that older brother Andrew has indicated a major sea change in energy and environmental policy. Sources inside the usually secretive organization indicate the elder Abadon spent a day last week at the California offices of Room for Tomorrow's CEO, Parker Parrish. After a protracted meeting at the RFT headquarters and a secondary meeting at an undisclosed location, Andrew Abadon has directed a major shift in holdings of energy stocks. Our sources indicate that he has already sold his personal share of the corporation's significant holdings in petroleum exploration but has not yet pulled his considerable assets from production and distribution."

"As might be expected, petro stocks were in disarray at the closing bell today." A pop-up box showed an aged stock video scene on the floor of the New York Stock Exchange with raised hands holding bid forms toward a harried floor worker. "In a statement, called in to the local Topeka affiliate, KSN-TV, Andrew Abadon stated, "I've just completed a 244-million-dollar directed investment in Room for Tomorrow's technology research efforts.

But more importantly, I've terminated Abadon Industries' relationships with a number of policy institutes that, if nothing else, promote a dominant reliance on petrochemicals, deny global warming, and generally promote a position that will put the United States in a back seat to China and Brazil as producers of sustainable energy."

The commentator went on to discuss speculated ramifications and introduced a panel of energy and climate experts. Parker, round-eyed, said. "244 million dollars?"

Carl burst out laughing, bent over. As his mirth took him further out of control, he grasped at tightened stomach muscles. To the wondering eyes of the gathering, he leaned back into the couch's cushions, a tear streak trickling down one side. "Yes, 243.7 million dollars. That's an approximation of the last transaction I made of Abadon's Cayman stash to our Cayman stash." He swiped at the tear streak, beaming broadly. "The crafty old bastard knew all along, or at least found out recently."

Parker put a hand on his shoulder, carefully. "Hon, are you OK?" Although lost in laughing, his face clearly looked pained.

"Yes, but my stomach wound still hurts like a bitch when I laugh."

"So how did they get the number right? Peck?" Brayton asked.

"Probably." Carl continued. "Or MacCauley, or Imbler. I've got to think that Peck had already found out, and his claims may have found other information that the Abadons were uncomfortable with him knowing. Like why all those millions were being secretly shifted around between legitimate banks in New York and Chicago and shadier locations in the islands and Switzerland." He shook his head. "No telling, really. But they fired him, and he was pissed and was leaving the country."

"Not before he was going to murder us!" Parker jabbed.

"Yeah, well, that didn't turn out well for him did it?" Grieg said. "That pile of crap is decomposing in a toxic hell." He paused, scratching at a three-day stubble. "Not so sure about that. We know he's not here, but I'm not so sure he's there, either." All heads turned toward him.

"Temporal paradox." He continued. "Change has happened and will be happening. The room is gone. CalStation is gone." His gaze scanned the room. "This future, the one facing us, never developed, or at least never deployed one of their time stations in that farmhouse. Maybe they put it somewhere else instead, if they put them anywhere, or hell, maybe they don't get invented, or the research goes to some other time." He shrugged. "It didn't go to where a future me would go to dig for clues in an underground data bank." He turned to his left to check Chu's reaction. To his and everyone's surprise, Chu was not there. Grieg swiveled his head toward the house to see if Chu was going for the head.

Simultaneously, everyone else noted Chu's absence. Carl jumped up first heading for the house. "Chu?" He disappeared into the house, but Carl's calls for their companion echoed in emptiness.

Joanie stood and reached for Grieg's hand, as if by holding it she could prevent whatever had happened to Chu from happening to Grieg. They all stood and faced the perplexed Carl as he stepped out on the poolside patio. "Well?" Parker asked.

"Chu is not in the house." He let that news settle in. A profoundly upset Carl looked off into the foliage, as if searching for words in the dense and confusing tapestry of bougainvillea branch and stem. He looked down, pointed at an empty table. "Chu's backpack is gone."

"What do you mean?"

"Where could he have —"

The tenuous nature of time hit them all in the gut.

"Chu!" Grieg's cry was loudest. "Solomon?" The realization, slow to solidify became apparent reality. Chu was gone. Solomon Chu, for whatever reason, could no longer exist in their current timeline. He sat back against the couch in gray ashen shock.

The party had gone from celebratory to sober in a moment. Grieg leaned forward, took a huge gulp to finish his drink, and slammed the tumbler down on the table. His eyes passed in turn from Joan to Carl then to Parker. His face a wreck in progress. Everyone shared the thought, when would he go? Grieg collapsed into the couch cushions. Joan wrapped his head in her arms and began to sob. The patio was silent except for the hum of distant traffic and a neighbor's sprinkler's staccato stk, stk, stk.

From down the hall, the sounds of a toilet flushing. "Hey," Chu called out as he walked back into the room to four astonished faces, "looks like someone died in here. Wot's up?"

~ ~ ~

A new reality settled in. The discussion went on for hours. Grieg got a little more drunk, but no one could blame him. His belief that he was securely safe in the 21st was shaken. Joan and Parker put him to bed amid consoling reassurances. Parker returned to the patio. Joan stayed behind and curled into the s-curves of Grieg's sleeping body, her arm coiled over his waist. Chu had collapsed into Carl's office couch and was sleeping the sleep of a margarita victim.

They were alone now, the celebration had turned to, to what? A new beginning for Grieg and Joanie? The end of a particular timeline for Tom Corliss and Odie Peck? Could an altered occurrence have any effect on their fates somewhere down a new timeline? Would Grieg and Chu's futures in the 21st ever be secure?

The effects of their political, economic, and technological manipulations had always been aimed at preventing a horrible and possibly avoidable human catastrophe. But the danger of playing with a future and its unknown outcomes had always been there. Parker and Carl settled into his living room's comfortable sofa, legs entwined as they had often done as they had nestled into her ocean view furniture. Carl reached out and put light pressure on her foot.

She pulled away from whatever reverie had captured her. She turned to face him. "Carl, what do we do now?"

"We keep moving." He let his foot caress her calf. "We have work to do. You have your list of projects to keep promoting. We maintain the same pressures on the same pressure points. Room for Tomorrow was always about long-term outcomes."

She nodded, slowly. Focusing on her ongoing mental to-do list, prioritizing. After a moment she looked up. "The Verts are history, right? They simply won't exist anymore, right?"

"Yeah, sure. I can live without that kind of excitement for a long time. Like the rest of my life." He locked eyes with her, almost completely sober now that they were discussing very real what-ifs. "Right now, the rest of my life is beginning to have a longer life expectancy attached to it."

"What about you? You really can't go back into the Abadon war chest for 'donations.'" She hooked air quotes with her fingers.

He turned his gaze out toward the lush foliage of the overgrown bougainvillea covering the patio walls. "There are still a few sources that I can tap, but I don't think I'm going to need to. Money follows money. Andrew just said he's thrown down millions of dollars on Room for Tomorrow. More will come. His buddies are going to give us a lot of thought. I think a lot more will come. You need to put a big push on the corporate sponsor list. And be sure that some of our investments have an equity line that will support our own growth."

"So that leaves you with some free time." Her head tilted, partially from exhaustion and partly in open question.

"Well, I've kind of morphed into project management anyway, and…I have an idea about some property investments."

"Yeah? Where?"

"There's a little valley north of here with some really cheap property. Place called Neenach. I've already put down an option to buy a couple of acres. I was thinking of developing some high-security data storage assets up there in an underground bunker." His hands described a large square in the air. "Fully wrapped in a Faraday cage against EMP attack." He let the beginnings of a grin dimple his right cheek. "I have the construction plans in this." He contorted to pull a data cube from his jacket pocket.

"I was thinking about creating a new company, an offshoot perhaps, of Room for Tomorrow, wholly owned and subsidized by RFT." He let the idea roll around, gel, gain substance. "We'd have some ground rules: no family archives, no pictures of cats, no porn, only secure corporate planning docs, research papers from university presses and think farms. I have a good name for it too, in deference to your roots in the North Florida woods. What do you think about Choctaw Cloud Storage Services?"

Parker's eyes opened wide. But her irises dilated, mirroring her critical evaluation. "Choctaw? Not Chick-a-Saw?"

He laughed, sighed, remembered. He mimed blowing her a kiss. "I don't see why not. We've already proved we can change the future."

She grinned, leaned over, kissed him. Then settled back into the overstuffed cushions, frowning in thought. "There's a thing." She searched gray matter for the reference. "One thing we haven't even thought about taking on yet that's going to be crucial."

"Jesus, what? We're already aiming a shotgun at the world's crying issues."

She snorted a humorless laugh. "Jesus *is* the target. Actually, his well-intentioned agents of sanctimony." She turned her stare toward a climbing vine of bougainvillea covering a trellis in the corner of his atrium. The plant had survived years of his neglect and still managed to blow her away with its shower of pink blossoms. In the two years since she had first seen it, it had outgrown its trellis and was climbing over and along the atrium wall. She took a deep breath and started in; "The church, the Catholic Church, the fundamentalist Baptists, Mormons, any one of a number of the holy rolling Christians of any stripe, or Muslims, Orthodox Jews, and who knows who else who hide behind a veil of dogma and dictate that a woman can't have the option of birth control, or set her own fate. In the Third World, women and girls, are still chattel, traded off at thirteen for a herd of goats."

"Here in the US, more and more states are plugging away at Roe vs. Wade, quoting scripture when it suits them to save every conceivable life, no matter if there was rape, incest, foolish fifteen-year-olds foolin' around, or any other mishap of timing."

Carl raised a hand. "Park, I get it. Maybe we can help there, but we have a lot on our plate. It's a stacked plate!"

"I'm sorry." She turned back to him; beginning to tear up. "I get so, so…"

"Passionate. I do get it. But let's work on the agenda we have for the moment, and we'll see how to work that in. It's important. Unchecked population growth is as challenging as anything we might try to change. Probably education is a best place to start. And lobbying won't hurt, here at home. Funding opposition candidates in red states, lots of places where money will help." He raised an eyebrow in question. "OK, we'll slide it from back burner to front and center as soon as we can."

Wow, she thought, *he is a lot more than eye candy.* "You amaze me, Mr. Reyes, you simply amaze me."

His smile reflected her own. She knew he might even be worth a third trip to the courthouse, someday. Then another cog tripped in her always moving mental gears.

"There's another thing." She searched gray matter for the reference. "Remember that we started the Verts because of the book by Abbey? *The Monkey Wrench Gang*?"

"Yeah. So?"

Well there's another one we need to pay heed to. A chronicle for, no." She screwed her lips in concentration. "Ahh. *A Canticle for Liebowitz*. It's by Walter Miller."

"OK, what's a canticle and who's Liebowitz?"

"It's a hymn, a song, not important. Liebowitz was an engineer whose writings and sketches became important to a civilization a thousand years or so after civilization rebooted from nuclear obliteration. But his mundane messages and sketches were elevated to mystical importance by some monastic types because, without any frame of reference, they couldn't understand a simple electrical circuit drawing or grocery list."

Carl looked up at the blank ceiling, musing.

"Right," he interrupted, understanding the reference and the warning, "we need to ensure that everything stored has directions for the directions. Math texts would need basic instructions for the symbology. We couldn't assume our symbol sets would be understandable." He thought about the future survivors in the 23rd. "So, our guys got lucky. They didn't deadfall all the way back to stone age illiteracy. But Liebowitz's followers just saw gibberish."

"Exactly. That's the point; we need to make sure our data bank would still make sense. We'll need a digital Rosetta Stone to include as many language roots and translations as possible."

His hand started to whirl in the air with excitement. "We'd need to include dictionaries, even literature so the cultural context is understandable."

She stood up, turned to face him, even more amazed. He rambled on. He'd gotten it so fast. Her smile spread, dimples deepening. "Carl, shut up!"

"What?"

Her eyes sparkled damply. "Let's go to bed." She leaned forward, blew a kiss.

He pulled her head to his, kissed her forehead, her cheek, her mouth.

She lifted away from him, rising from the couch. She bent, and grasping for his hand, pulled him to his feet.

"Come on. I need grounding, then sleep. Plenty of time to save the future, Tomorrow."

The End

If you've enjoyed this book, please go to Amazon.com and provide a review. Five stars are especially appreciated if it made you think about this small blue planet as a very special place. It's endangered, and you know it. As a self-published author, I need all the help I can get and your review will certainly help. Thank you, I hope you enjoyed the journey.

EPILOGUE

I'm providing this list of resources knowing that in future editions I will have to ensure that the links are still valid. The internet and its many layers are fluid in time. Error 404 is all too common. The subject lines will change with time, but the need will not change. There is a world that needs saving. Maybe that's too pretentious? Saving the world?

I don't think there's a much larger issue facing the BIG WE, the All of Us, the Human Gestalt, than finding a way to keep the excesses and successes of the 20th century and its lingering momentum into the 21st century from killing us all off. Who is to blame for the mess, our proximity to the precipice? Almost every one of us shares some of the bill with our personal carbon footprint, our demands on the grid. By simply being, we move the collective whole ever upward toward the asymptotic limit that our dear Mother Earth cannot bear.

With increasing alarm, scientists have been proclaiming that in a hundred, or eighty, or fifty, or ten years, our continuing increased production of greenhouse gasses will push complex ecosystems and climatic systems to their limits, to their individual asymptotes. When will enough of the Canadian and Siberian tundras lose their permafrost to become methane factories that pull the final trigger? When will there not be enough snowfall to keep glaciers alive, when will Greenland lose its ice sheet, or Antarctica its vast store of frozen water? When will our super crops no longer produce seed stock in growing seasons too warm to produce viable seed?

This *is* a book of fiction, speculative fiction, with just enough science to call itself, science fiction. Now that you've finished Parker Parrish's journey, please explore the references below. It is my intent that they will be updated and live on in the digital versions and paper reprints of this book. They are the websites of the numerous organizations that have seen the challenges and have been working toward solutions big and small for decades. If you have the electronic version of this book, you can simply tap the URL. If the address does not work, try the root URL for the site and browse for updated articles.   Updated 10/04/2021

RESOURCES

• Reforestation:
https://news.globallandscapesforum.org/tag/landscape-restoration/
https://metro.co.uk/2018/08/08/man-turned-desert-forest-planting-treeevery-day-40-years-7814241/
http://www.treenm.com/why-plant-trees-in-the-desert/
https://www.greenpeace.org/usa/forests/
https://edenprojects.org/the-problem-and-the-solution/
https://www.alivingtribute.org/

• Tidal/Wave Power:
https://www.nationalgeographic.org/encyclopedia/tidal-energy/
https://www.boem.gov/renewable-energy/renewable-energy-program-overview
https://www.planete-energies.com/en/medias/close/tidal-energy
https://www.eia.gov/energyexplained/hydropower/tidal-power.php

• Ocean Current Power
https://tethys.pnnl.gov/marine-renewable-energy
https://www.youtube.com/watch?v=5sBs3QUErWM
https://www.planete-energies.com/en/medias/close/ocean-current-energy
https://www.power-technology.com/features/featuretidal-power-floridas-ocean-current-potential-4379164/
https://simecatlantis.com/projects/
https://snmrec.fau.edu/ocean-energy/ocean-energy-fl-straits.html

• Loss of Permafrost:
https://www.nationalgeographic.com/environment/article/thawing-permafrost-forces-denali-national-park-to-reimagine-its-future
https://e360.yale.edu/features/how-melting-permafrost-is-beginning-to-transform-the-arctic
https://www.canadiangeographic.ca/article/arctic-permafrost-thawing-heres-what-means-canadas-north-and-world

https://www.thearcticinstitute.org/permafrost-thaw-warming-world-arctic-institute-permafrost-series-fall-winter-2020/
▪ Loss of Sea Ice:
https://www.sciencedaily.com/releases/2019/11/191112114009.htm
https://www.whoi.edu/press-room/news-release/greenland-ice-sheet-melt-off-the-charts-compared-with-past-four-centuries/
https://climate.nasa.gov/vital-signs/arctic-sea-ice/

▪ Solar & Battery Research:
https://pubs.acs.org/doi/abs/10.1021/am2017909
https://www.sciencedaily.com/releases/2019/04/190425104251.htm
https://www.nature.com/subjects/batteries
https://www.energy.gov/eere/solar/photovoltaics-research-and development
https://www.nrel.gov/news/program/2019/breakthrough-method-creating-solar-cell-scientists-prove-impossible-is-not.html

▪ Carbon Research:
https://arpa-e.energy.gov/technologies/projects/graphene-based-supercapacitors
https://www.ramacocarbon.com/
https://news.berkeley.edu/2018/08/13/long-sought-carbon-structure-joins-graphene-fullerene-family/
(and you thought I made this stuff up)

▪ Educating Muslim girls:
https://www.irfi.org/articles/articles_501_550/educating_muslim_girls.htm
https://www.whyislam.org/social-ties-2/the-importance-of-girls-education-in-islam/
https://onlinelibrary.wiley.com/doi/full/10.1111/padr.12142
https://onlinelibrary.wiley.com/doi/full/10.1111/padr.1212

▪ Initiatives:
https://www.un.org/en/climatechange/climate-action-coalitions

- The Non-Fiction, really-doing-the-job-publicly-funded-and-usually-not-for-profit people really doing the work:

Beyond Carbon   https://www.beyondcarbon.org/
Center for International Environmental Law   https://www.ciel.org
The Conservation Fund   https://www.conservationfund.org
Earth Justice   https://earthjustice.org
Environmental Defense Fund   https://www.edf.org/
Defenders of Wildlife   https://defenders.org
Green Peace   https://www.greenpeace.org
League of Conservation Voters   https://www.lcv.org/
National Parks Conservation Association   https://www.npca.org
National Resources Defense Council   https://www.nrdc.org/
The Population Connection   https://www.populationconnection.org
Sierra Club   https://www.sierraclub.org
Society of Environmental Journalists   www.sej.org/
The Nature Conservancy   https://www.nature.org/en-us/
Friends of the Earth   https://foe.org/
The Trust for Public Land   https://www.tpl.org/
Union of Concerned Scientists
https://blog.ucsusa.org/tag/environmental-protection
United Nations https://www.un.org/en/climatechange/climate-action-coalitions
The Wilderness Society   https://www.wilderness.org/
World Wildlife Fund   https://www.worldwildlife.org/initiatives/climate

- And the inspiration of this book: Lester Brown
   http://www.earth-policy.org
   https://en.wikipedia.org/wiki/Lester_R._Brown

The website is posted and no longer updated, but this book is now a free download:
   http://www.earthpolicy.org/images/uploads/book_files/pb4book.pdf

I'm sure there are dozens or more organizations dedicated to making a difference in our, in your, future. If I haven't listed one that you feel is worthy of inclusion, email me at bruce.ballister.com

- And since I'm listing inspirations:

Edward Abbey's *Monkey Wrench Gang* is available here;
https://tinyurl.com/466mfcn9

Made in the USA
Columbia, SC
26 January 2022

54269289R00211